Thomas Seebeck:
His Life and Science

Daniel Foty

Copyright © 2018 by Daniel Foty.
All Rights Reserved.

Also by Daniel Foty:

Moore's Law: The Untold Story (2017)

The Rise and Fall of Classical Athens: A Short History (2018)

Table of Contents

Preface vii
Chapter 1: A Short Biographical Sketch 1
Chapter 2: Matter, Heat, and Light Before Seebeck 8
Chapter 3: The Scientific Work 31
Chapter 4: The Legacy 45

Preface

"Thomas Seebeck" is a name that is familiar to most students of physics; this is because Seebeck, in the early part of the 19th century, performed early and fundamental experimental studies of the relationship between heat and electromagnetism in the behavior of materials – and is thus remembered for these early studies in "thermoelectricity."

But is there more to the story?

It turns out that there *is* more – much more.

In 2010, I had the good fortune to be invited to give one of two lectures at the Baltic Electronics Conference on the life and scientific achievements of Thomas Seebeck, as part of events marking the renaming of the Department of Electronics at the Tallinn University of Technology – which was being renamed "The Thomas Johann Seebeck Department of Electronics."

It is little known that Thomas Johann Seebeck was born in Reval – which is the once-used name of what is today, Tallinn, Estonia. Seebeck grew up in Reval/Tallinn, and lived there until he was 17 years old. Being born into a prosperous merchant family, Seebeck's youth was spent in a fine house that faced (and still does) onto the main square of the old town. Given his family's means, after he completed his youthful education, he was sent off to Germany to further his education – never to return.

During his time in Germany, while working on (and eventually completing) his medical studies, he developed an intense interest in the various new findings that were opening up in many branches of physical science. It quickly became obvious that Seebeck was both scientifically talented *and* experimentally talented; he made important contributions to numerous aspects of basic science.

In addition to his now-well-known studies of thermoelectricity (which he actually carried out rather late in his life), Seebeck also made fundamental contributions in optics (birefringence) and color that provided the foundation for what would eventually become color photography, along with fundamental contributions in the optical behavior of organic and biological molecules that eventually led to biochemistry and molecular genetics. Interestingly and in addition, Seebeck's work in thermoelectricity was to provide an important tool for the study of the behavior of electronic materials – which eventually provided early hints of the potential of semiconductor materials (and their technological potential), and also provided insight that superconductivity (when discovered) represented something that was a fundamental departure from all prior known behavior in the field of solid state physics.

For the two aforementioned lectures, my old colleague and friend from the Tallinn University of Technology, Professor Enn Velmre, was tasked with describing Seebeck's biographical details along with details of Seebeck's experimental studies – reflecting a paper that Prof. Velmre had published in 2007. For my part, I took on the task of examining larger-scale issues associated with Thomas Seebeck's life and work – the history of knowledge in fields in which Seebeck worked (prior to Seebeck), Seebeck's work, his contemporaries (scientific and otherwise), and the legacy of what followed from Seebeck's work in the several fields where he made important contributions.

It is these aspects of Thomas Seebeck's life and work that are the subject of this short book. The first chapter provides a brief biographical sketch of the life of Thomas Seebeck – but this book is not a biography. The experimental studies carried out by Seebeck during his lifetime are considered in some detail – but, in order to provide context for the importance of that work, the prior knowledge (gained over centuries) is considered first, and the wide-ranging (and little-recognized) implications that followed from Seebeck's work are considered in detail.

■■

In converting my 2010 lecture into this book, I must thank several colleagues and friends for their help with bringing the book to fruition. First, I must thank Prof. Enn Velmre of the Tallinn University of Technology for his help with the basis of the book; Prof. Velmre was kind enough to allow me to use his 2007 paper on Seebeck's life history as the basis for much of the biographical sketch that is presented in Chapter 1; Prof. Velmre also provided several important clarifications and corrections to that chapter, and also provided one of the photos. Also in Tallinn, I must thank Prof. Peeter Ellervee of the Tallinn University of Technology for providing encouragement for the idea of turning the lecture into a book, for reviewing parts of the manuscript, and for continuing to believe that the bi-annual Baltic Electronics Conference cannot be held without my articipation. Finally in Tallinn, I must thank Mari-Ann Kelam for providing one of the photos.

In addition, I must thank Prof. Bart Van Zeghbroeck of the University of Colorado for providing the data sets of the Peltier coefficient of n- and p-doped silicon over temperature – behavior which Seebeck's techniques were able to find, and which (as mentioned above) provided early hints of the technological potential of semiconductor materials (such as silicon). I also must thank Tom Weller for permission to use two figures (one serious, the other humorous) from his 1985 classic "Science Made Stupid." This delightful book is (sadly) out of print, but it can be found online.

■■

With regard to the figures in this book, I have endeavored as much as possible to use figures that are clearly known to be public domain, created by me, or used with the appropriate permissions. However, there are a few figures which are of uncertain provenance but which are simply too good and too unique to readily replace. Hopefully no copyrights or courtesies have been violated with respect to those figures – but if this turns out to be the case, I will endeavor to repair the problems.

Fletcher, Vermont
September 2018

Chapter 1: A Short Biographical Sketch

As was noted in the Preface, this book is not intended as a biography of Thomas Seebeck; instead, the goal is to examine the background, context, and implications of Seebeck's work – and thus to more-completely understand his contributions to scientific and technical knowledge.

However, it is appropriate to first provide a short biographical sketch – of the life of Thomas Seebeck. Following this, some temporal context will be provided – with respect to the period in which Seebeck lived, and with respect to several of his contemporaries (and near-contemporaries).

A Short Biographical Sketch

(As a source for this section, I have used a short paper on the subject of Thomas Seebeck's life and work: E. Velmre, "Thomas Johann Seebeck (1770 – 1831)," Proceedings of the Estonian Academy of Science and Engineering, vol. 13, no. 4, pp. 276 – 282 (2007); this paper can be readily found online. I am grateful to Prof. Velmre and his colleagues at the Tallinn University of Technology for their permission to use this paper as the source for the short biographical sketch which follows.)

The roots of the Seebeck family in Estonia are connected to the time when Estonia was part of the Swedish Empire. The oldest identifiable ancestor of Thomas Seebeck in Estonia is Klaus Seebeck – who, in the early 1600s, came to what is now Estonia to serve as the commander of the Swedish eastern border fort at Jamo (also known during that period as Jama, Jam, or Jamburg), which today is Kingisepp, Russia. Klaus Seebeck's descendants remained in what is now Estonia, and some eventually migrated to Reval (the original medieval name for what today is Tallinn) – where Thomas Seebeck's immediate lineage became a successful merchant family. In the early part of the eighteenth century, following the conclusion of the Great Northern War, control of what is now Estonia passed from the Swedish Empire to the Russian Empire; however, as was common at the time, the Seebeck family remained in Reval and continued their prosperous merchant activities.

Thomas Johann Seebeck was born on April 9, 1770, the son of Johann Christoph Seebeck and Gerdrutha (Lohmann) Seebeck, in Tallinn/Reval;

as noted above, at that time what is now Estonia was part of the Russian Empire. His merchant family was wealthy enough to own (and inhabit) a comfortable house which faced directly onto the main (Old Market) square of Reval – as was common during this period in many of the commerce-centric cities of eastern Europe, it was the wealthiest and most-successful merchant families who owned and lived in the houses which faced the main square of the town.

The Seebeck house is still standing in Tallinn, facing the main square in the lower town. Today, it is a high-end women's clothing store.

Figure 1.1. The Seebeck family house on the main square of Tallinn, the Old Market Square. (Photo courtesy of Mari-Ann Kelam; used with permission.)

Figure 1.2. Another view of the Seebeck family house on the main square of Tallinn, the Old Market Square. (Photo courtesy of Prof. Enn Velmre; used with permission.)

In the spring of 1787, Seebeck graduated from the Reval Imperial Grammar School – which still exists today in Tallinn, and is now known as the Tallinn Gustav Adolf Grammar School. At that time, what is now Germany was regarded as the highest center of scientific and technical excellence in Europe; thus, following his graduation, Seebeck left Tallinn for further education in Berlin and Göttingen. He enrolled at Göttingen University, and in 1792 completed his education in medicine and surgery; he received the degree of Dr. Med., and thus was qualified to be a practicing medical doctor.

Figure 1.3. A plaque which today adorns the side of the former Seebeck family house on the Old Market Square in Tallinn. (Photo by the author.)

However, during his time in Germany, Seebeck had become taken with the new and interesting developments that were occurring in the basic sciences. With his medical qualifications in hand, he took up residence in Bayreuth. However, medicine was not to be the focus of his life. His father had died in 1786, leaving Seebeck with a considerable inheritance; this allowed him to eschew a medical career – and to instead devote himself to his scientific interests (particularly in physics).

By 1795, Seebeck had married Juliane Boye, and their first child soon arrived. Sometime after that, he returned to Göttingen University, and earned a doctoral degree in medicine in 1802. Following his receipt of this degree, Seebeck (and his wife and children) moved to Jena.

Seebeck's years in Jena proved to be the most developmentally-important of his life – as the European Enlightenment of the eighteenth century reached its full flower during the nineteenth century. Whether by accident or design, Seebeck's presence in Jena put him in one of the centers of a variety of activities associated with this flowering – which encompassed not only science and engineering, but also philosophy, literature, and other fields. Of the many now-famous people with whom Seebeck became associated during his time in Jena, his life-long friendships with the philosopher Georg Wilhelm Friedrich Hegel, the explorer Alexander von Humboldt, and the poet/polymath Johann Wolfgang von Goethe were the most notable; as will be detailed in Chapter 3, Seebeck and Goethe worked extensively together in the field of optics and color.

In 1810 Seebeck returned to Bayreuth, then moved on to Nuremberg. In 1818, while he was living in Nuremberg, he was elected a corresponding member of the Prussian Academy of Sciences in Berlin. Shortly thereafter, in January 1819, Seebeck was named an academician (full member) of the Prussian Academy of Sciences; this prompted him to relocate from Nuremberg to Berlin during the summer of 1819. Even though he had yet to do the work for which he is now best-known (his work on thermoelectricity), his scientific reputation was already more than high enough to merit these honors and the appointment to the Academy of Sciences. It was in Berlin that Seebeck was to carry out (and report on) his now-famous work on thermoelectricity; from 1820 to 1822, Seebeck gave five presentations on this work at the Academy, with the first publication appearing in 1822.

Unfortunately, Seebeck's health began to deteriorate badly in 1823; his ability to work (and to report on that work) declined along with his health. Thomas Seebeck died in Berlin on December 10, 1831.

Thomas Seebeck and His Contemporaries

As was just described, Thomas Seebeck's life (and his most productive years) overlapped with those of a number of remarkable thinkers and doers of the same period. One such nearly-perfectly-matching overlap is worth noting on its own.

b. Tallinn, 9 April 1770
d. Berlin, 10 December 1831

b. Bonn, 16 December 1770
d. Vienna, 26 March 1827

Figure 1.4. Thomas Seebeck and Ludwig van Beethoven.

This is rather remarkable – and also serves to place Seebeck's life in the context of the period in which he lived.

It is also interesting to expand this comparison – to include several other notable personalities (scientific and otherwise) of about the same period of time:

Thomas Johann Seebeck: 1770 – 1831
Ludwig van Beethoven: 1770 – 1827
Joseph Fourier: 1768 – 1830
Napoléon Bonaparte: 1769 – 1821
Johann Wolfgang von Goethe: 1749 – 1832
Thomas Jefferson: 1743 – 1826
Jean-Charles Peltier: 1785 – 1845

This list provides further context with respect to the period of time in which Seebeck lived and worked.

As a final piece of context, it is worthwhile to examine a chart depicting the lifespans of the various scientific (and other) personalities who will appear in this book; this particular chart covers the period and personalities beginning with Thomas Newcomen's development of the first practical steam engine, and ends with the codification of the laws of electromagnetism by James Clerk Maxwell.

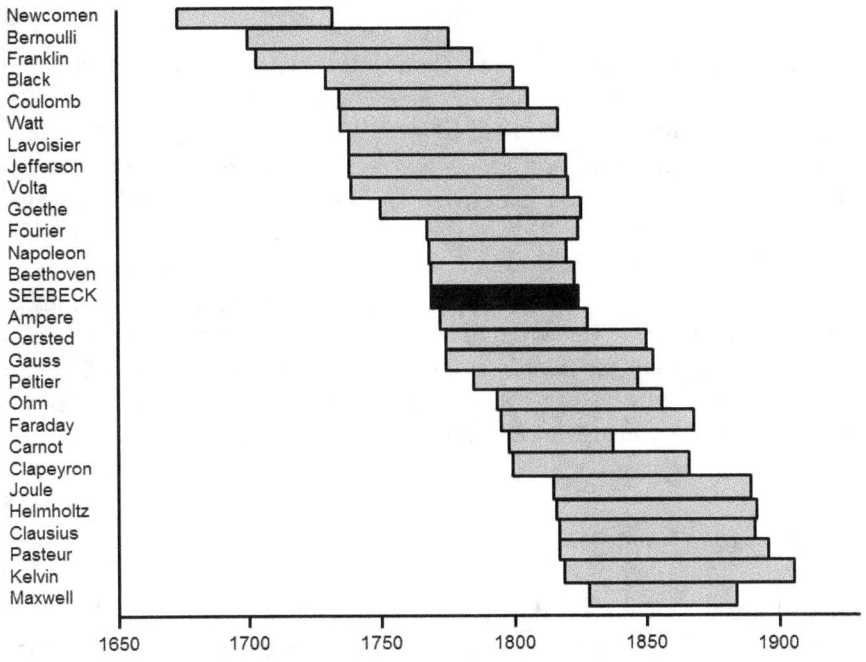

Figure 1.5. A chart depicting the lifespans of Thomas Seebeck and various other (mostly-scientific) personalities; the scientific personalities are mentioned later in this book – as scientists who provided work upon which Seebeck built, and as scientists who built upon Seebeck's work.

Along with the short biographical sketch, the description of Seebeck's contemporaries (and near-contemporaries) has hopefully provided insight (and context) into the times and world in which Thomas Seebeck lived – and performed his work.

Chapter 2: Matter, Heat, and Light Before Seebeck

Since time immemorial, humans (and our even-older ancestors) have looked at the world around them – and contemplated what they saw of things such as matter, heat, and light. While knowledge was limited, practical experience allowed these features of the natural world to be manipulated and put to beneficial use over hundreds of thousands of years.

Scientific knowledge grew slowly – and accumulated just as slowly. Such knowledge required the observation that the natural world behaves rationally – that is, the behavior of matter, heat, and light is not arbitrary, but follows clear-and-repeatable rules. If the world is rational rather than arbitrary, then it can be subject to study – in order to find the rules that govern the natural world, and to put those rules to practical use.

The Classical Greek Era

During the classical era (c. the seventh century B.C. through c. the fourth century B.C.), Greek thinkers ("philosophers") began to notice that there appeared to be rational order in the natural world, and began to try to understand the underlying principles. While these thinkers were not "scientists" as we would now understand the term, they did lay the foundation for the branch of philosophy which considered the behavior of the natural world – "natural philosophy," which is the ancestor of contemporary science. This idea is retained even today in the name for the most advanced degree in a subject (including the sciences) – "Ph.D.," which is the Latin form of the phrase "Doctor of Philosophy."

One of the main differences between contemporary "science" and classical Greek "natural philosophy" is that classical Greek thinkers made some simple observations about the world around them – and then extrapolated those observations into convenient universality. One prominent example is the observation of things being in flux rather than being static – leading to the conclusion that *everything* is in flux, and therefore the further conclusion that *any* seeming form of stasis is actually an illusion (with the flux being present in there somewhere, if one looks hard enough). The classical Greeks also had abiding beliefs in the notion (based on the idea of "flux") that everything in the world runs in cycles, and also that everything in the world will actually exist in symmetric form (and in the most symmetric forms possible). This led the Greek natural philosophers to fill in gaps in their knowledge by imposing these principles without concomitant observation – and to believe that they could figure out answers via simply thinking about confounding problems.

Despite these shortcomings, the Greek natural philosophers did accomplish a great deal in developing an understanding of the natural world around them – and they did indeed lay the foundation for present-day science and technology.

Perhaps the original Greek natural philosopher was Heraclitus, who lived in Ephesus (on the coast of Asia Minor) around 500 B.C.

Figure 2.1. Heraclitus of Ephesus.

It was Heraclitus who first described the idea of flux that was detailed above – that "all things are flowing" in one way or another. Heraclitus' most famous philosophical observation along those lines was that one can never step in the same river twice – since by the second time new water has flowed in to replace the old, and the person in question will also have changed by the second time in the river.

Heraclitus' main contribution to things that were closer to science was his postulation that the wide variety of matter that one sees in the world around us is actually not as wide in variety as it appears to be. Heraclitus postulated that all forms of matter are actually composed of combinations of only three basic elements – earth, fire, and water. This provided the important idea that the world around us is not as complex as it at first appears to be – that instead, the world is built up into great complexity from much simpler basic building-blocks; this idea is actually one of the foundation-stones of science and the scientific method.

Heraclitus' ideas were expanded upon by a natural philosopher living at what was pretty much the other end of the classical Greek world – Empedocles, who lived in the city of Acragas in Sicily.

Figure 2.2. Empedocles of Acragas.

The most-enduring idea of Empedocles was his expansion of the Heraclitus' three constituent elements of matter to four – earth, *air*, fire, and water.

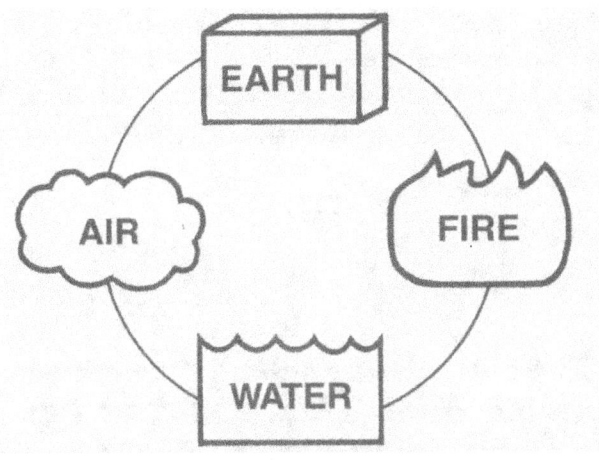

Figure 2.3. Empedocles' four basic constituents of matter. (Copyright 1985, Thomas W. Weller; used with permission.)

In classical Greek fashion, Empedocles' construct had the advantage (over Heraclitus' construct) of being more symmetric and orderly. It also had the advantage of postulating that the four elements could be combined to produce descriptions of other aspects of the natural world – where the classical Greeks had also developed a belief in opposite aspects of reality, such as hot-vs.-cold and wet-vs.-dry, and so forth.

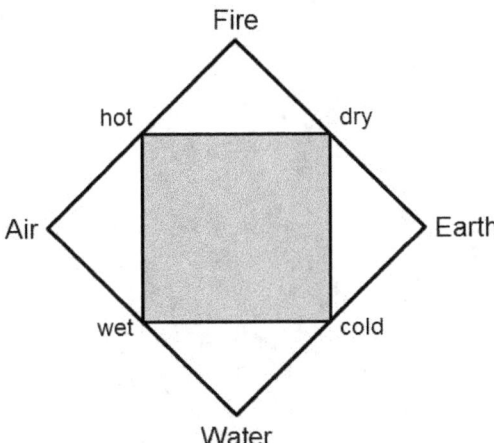

Figure 2.4. The system of the four elements, where combinations of the elements reflect observed opposite-and-opposing aspects of the natural world.

Of particular note, Empedocles' scheme (as expressed in Figure 2.4) can be extended into more practical areas.

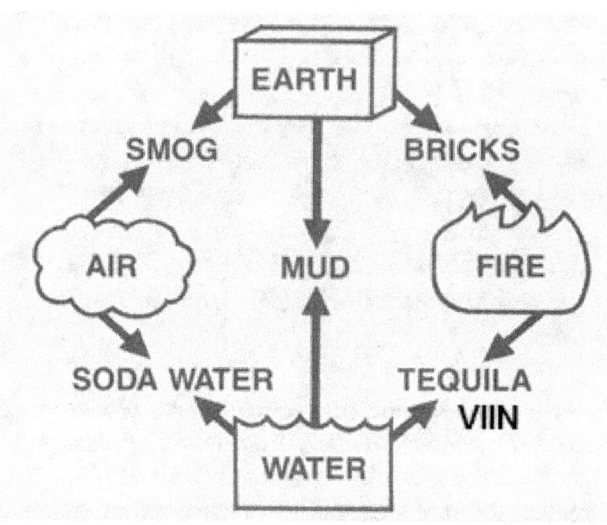

Figure 2.5. A practical update to the scheme proposed by Empedocles as embodied in Figure 2.4. (Copyright 1985, Thomas W. Weller; used with permission.)

And also, of course, there was Democritus, who was born at Abdera in Thrace.

Figure 2.6. Democritus of Abdera.

Although Democritus was originally from the seemingly-distance backwater region of Thrace, he traveled widely and wrote on a great variety of subjects. He was also known as "the laughing philosopher" – and his good humor apparently paid off handsomely, as he reportedly lived for 90 years (at a time when such a thing was virtually impossible).

Democritus' main contribution to Greek thought was an extension of the idea of the four elements – to the idea that at the base of all matter were individual aspects of that matter which could be neither divided nor reduced any further. Democritus' name for these indivisible base constituents was taken from the Greek word for "indivisible" itself – *atomos*.

Obviously, from that word comes the concept of "the atom" – as the most basic, indivisible constituent of matter. The name came to be applied to the most indivisible aspect of matter that still remains-and-retains what the original "matter" is; for example, if one keeps dividing a piece of copper into two pieces repeatedly, at some point that division will reach a minimum level of something that can still be identified as "copper" – a single copper atom. On the other hand, this assignment of "atom" (in the pure sense of "indivisible") turned out to be incorrect, as the atom can be further subdivided into basic particles (protons, neutrons, electrons) – and even those particles can be further sub-divided. But this misassignment cannot be blamed on Democritus – as his philosophical description was later used to define an "atom."

As a classical Greek philosopher, Democritus was not engaging in a scientific inquiry into the state of matter (as we would define it). He was philosophizing about how the state of the world must be – and in this case, he was essentially correct.

These classical Greek ideas seem quirky to modern eyes and minds; however, they represented a start – on the idea that the world was rational and comprehensible, and that the great complexity of the natural world could be created from a small set of basic building-blocks.

The Hellenistic Era

Eventually, the classical Greek era gave way to the Hellenistic Era (the late fourth century B.C. to the late first century B.C.). Due to the widespread conquests of Alexander the Great around the eastern Mediterranean and deep into Asia, classical Greek ideas, the Greek language, and Greek settlers spread all around what had been Alexander's world. Classical Greek culture and ideas went to all of these places, producing many new developments over a shockingly-large (by classical Greek standards) geographical area.

One of the more fertile fields of the Hellenistic era appeared in the Egyptian kingdom that was created out of the remnants of Alexander's empire – by his general Ptolemy, who became Pharoah and founded a dynasty that would last for nearly three centuries.

The Ptolemaic center of learning and natural philosophy grew up around Ptolemy's (Alexander-founded) capital of Alexandria, on the Mediterranean coast of Egypt, at the western edge of the sprawling Nile River delta. Thinkers (including many natural philosophers) flocked to Alexandria; Ptolemy and his successors encouraged this inflow, and also build up a vast collection of manuscripts that was stored in the famous Library of Alexandria.

One of the more interesting developments at Alexandria was one produced by Eratosthenes in the third century B.C. By observing the differences in shadows cast at various places along the (north-south) Nile River valley, Eratosthenes deduced that the Earth is a sphere. Further, by using the differences in the lengths of those shadows, Eratosthenes used simple geometry to calculate the diameter of the Earth itself – a result which was actually fairly accurate, and which was not superceded until the Renaissance.

Of particular note with respect to matter, heat, and light, one of the more interesting developments was produced in Alexandria by Heron.

Figure 2.7. Heron of Alexandria.

Around 50 B.C., Heron actually built the first known steam engine.

Figure 2.8. Heron's simple "steam engine."

This device is a "steam engine" in the sense that water is heated into steam and the steam exits the vents – as a result, the sphere spins around the provided axle. This "steam engine" is interesting, but as configured it does nothing practical.

Somewhat strangely, Heron's interesting but effectively-useless "steam engine" provided the end-mark of knowledge and learning of the Classical and Hellenistic Eras.

The Eclipse

Heron's steam engine of c. 50 B.C. was created near the end of the Hellenistic Era; this era "formally" ended in 31 B.C., when Octavian utterly defeated the fleet of Mark Anthony and Cleopatra at Actium on the Greek coast; after Actium, Octavian deposed the last monarch of the Ptolemaic dynasty (the famous Cleopatra), and reduced Egypt to a Roman province. Egypt was the last-remaining nominally-independent state of the old Hellenistic world; piece by piece, the Romans had taken over all of the other Hellenistic states around the eastern Mediterranean.

One of the great mysteries of the Hellenistic and Roman eras is the stagnation of science and technology – and the dearth of new ideas. At the time that Heron developed his "steam engine," the people of the Greek and Hellenistic world had long been familiar with basic mechanical devices; gears and cogs had long been readily fabricated from wood, and had been put to use for practical purposes (such as for the grinding of grain). By the time that Heron created his "steam engine," the Hellenistic world had all of the pieces that were necessary to put together a genuinely-practical steam engine.

But for some reason, no one thought to pull all of those pieces together into such a practical steam engine – one that could power various "engines" that could grind grain, propel ships, etc. This is actually not unusual in human history – for a society to have "all of the pieces" available, but to not put them together into a practical technology. For example, in the pre-Colombian Americas, wheels appeared on children's toys – but were never put to use for transportation purposes or other practical uses; in addition, although the ability to smelt melt ore was well-known and common, the development of alloying (which provides metal implements of practical strength) never occurred – and military technology, even of the great pre-Colombian empires, never exceeded the use of stone weapons.

What exactly happened in the Hellenistic world is unclear. In both the Hellenistic and Roman worlds, warfare was common – and warfare led to large numbers of enemy captives (captured troops who had surrendered), and it was common for those captives to be sold into slavery as a way of deferring the enormous expenses associated with military campaigns. With the vast availability of inexpensive muscle power (both human and animal), it is possible that there simply was little impetus to develop technological power sources – muscle power could forge metal and mill grain, ships could be powered by sails or by rowers (during the Classical, Hellenistic, and Roman eras, most "sailors" were actually rowers; this was even the case for the Vikings!), and armies moved around mostly on foot (infantry troops moved around mostly by "marching" – i.e., by walking from point A to point B) until well into the twentieth century.

The Hellenistic world also suffered greatly from what today we would call "academic snobbery." The "natural philosophers" at Alexandria (in particular) regarded themselves as being "philosophers" – thinkers who were well above the practical realities of daily life, and who would have been horrified at the notion of applying their "philosophizing" to practical purposes.

The development of science and technology in the Greek and Roman worlds stagnated. There were some Roman thinkers, such as Pliny the Elder (who deduced that amber is petrified pine resin, and who also noted some early aspects of electrostatics and electrodynamics), but a technologically-stagnant world of muscle-power largely prevailed. For centuries, most of the interesting scientific and technological developments in the world occurred in China.

But in seventeenth century, this began to change radically.

The "Enlightenment" – What is Heat?

For this condensed background story of matter, light, and heat, the story can now skip ahead to 1600.

The story picks up in England, with Francis Bacon.

Figure 2.9. Francis Bacon

On a larger scale, Bacon is often regarded as the father of what we now call "the scientific method." As was described earlier, the natural philosophers of the classical Greek period were thinkers – but their approach to "science" left quite a bit to be desired. In particular, the Greek thinkers had basically a metaphysical view – of the need for the universality of particular organizing phenomena (such as a belief in the need for symmetry, as well as a belief in the need for flux and for cycles); as a result, the Greek thinkers would try to impose these conditions as *necessary* (even when they were not actually observed – that they must be in there *somewhere*), and would often try to arrive at "conclusions of fact" entirely by thinking about explanations for what they saw. Bacon provided the framework for the idea of scientific inquiry – of how facts have to be what they are, and that postulated explanations are hypotheses which can be tested (and possibly thus discarded) via further observations and controlled experiments.

In terms of his contribution to the general problem of matter, light, and heat, Bacon was able to deduce that "heat" is actually motion of some sort.

Also in England, in 1712, Thomas Newcomen came up with a practical use for heat – when harnessed into a mechanical apparatus.

Figure 2.10. Thomas Newcomen.

At that time, the first flourishes of industrialization were appearing around England. While wood had been the traditional source (for millennia) for producing heat via combustion, it was known that coal is actually a much better source of heat. England has plentiful supplies of coal under its soil – but mining such coal quickly led to a practical problem – mines would easily flood, rendering them useless. Newcomen developed the first practical steam engine, as an apparatus for pumping water out of mines.

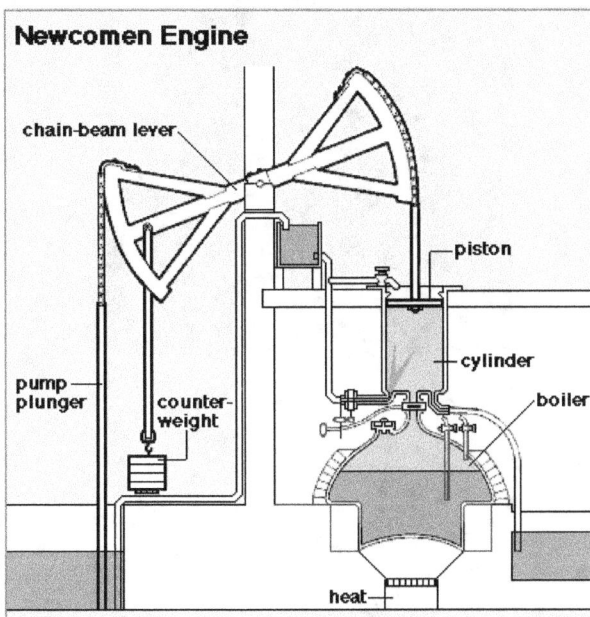

Figure 2.11. The "Newcomen engine," the first practical steam engine.

Newcomen's engine was a practical triumph, as it allowed the mining of coal to be pursued on a large and reliable scale. But this steam engine was also a scientific triumph; it converted heat into steam (from water), the steam was used to cause motion, and the motion did the practical work of pumping water up and out of a mine – writ large, it converted heat into work. This was a result obtained not from "theory," but from practical observation; in contrast to the notion that "theory" always precedes "practice," the scientific reality is actually closer to that of the steam engine – that improvements in practice lead to improvements in theory, while the improvements in theory lead to improvements in practice, with one continually being improved by the other.

Furthering Bacon's deduction that heat was actually a manifestation of motion, in 1738 the Swiss scientist and mathematician Daniel Bernoulli developed the kinetic theory of gasses.

Figure 2.12. Daniel Bernoulli.

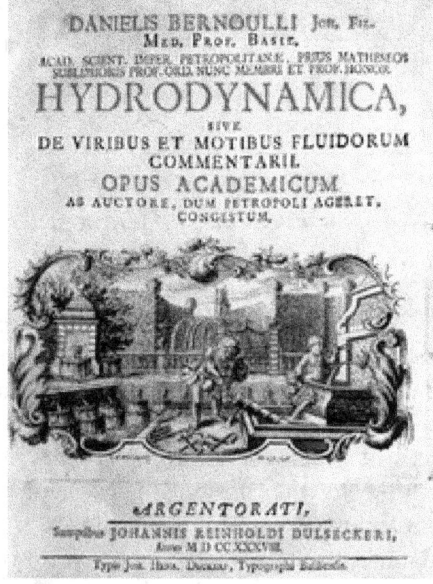

Figure 2.13. The cover of Daniel Bernoulli's 1738 book.

Bernoulli based his work on the notion that a gas is comprised of individual constituent particles; he then demonstrated that when the gas becomes "hotter," this is as a result of the particles of the gas taking on added energy and moving more quickly.

Meanwhile back in England, in 1761 Joseph Black made some interesting observations.

Figure 2.14. Joseph Black

Black observed the simple physical phenomenon of ice melting, and noticed something "strange" – that when heat is applied to ice, the ice melts into water... but during this melting, the temperature of the ice-and-water does not change. Black had noticed that the behavior of phase transitions was something unique.

Also in England, in 1776, James Watt introduced his improvements on the Newcomen steam engine.

Figure 2.15. James Watt.

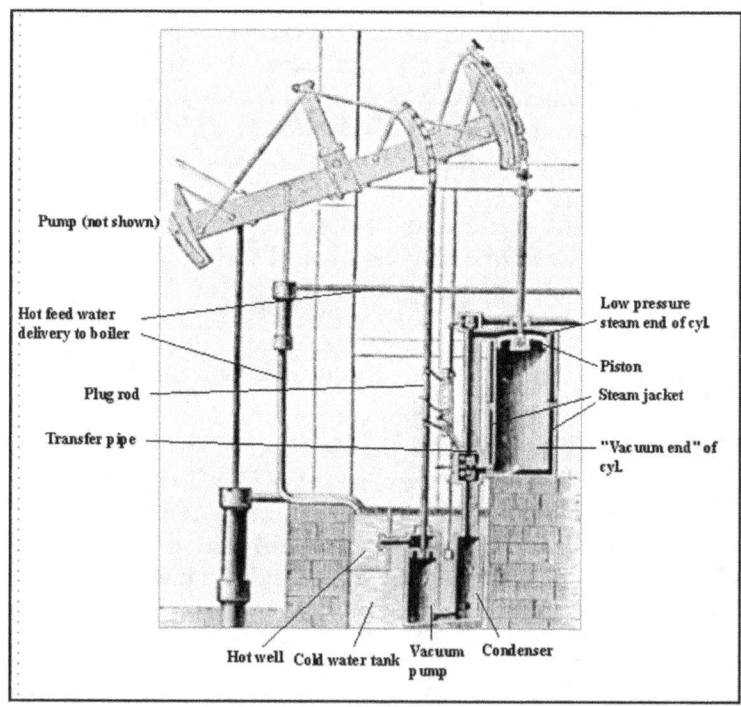

Figure 2.16. Watt's improved steam engine.

This steam engine was much more efficient and effective than Newcomen's original steam engine, and it proved to be much more effective in the task for which Newcomen had developed his steam engine – the pumping of water out of mines. Watt's steam engine was so effective that it made the large-scale mining of coal much more efficient – and it was this plentiful supply of inexpensive coal that caused the Industrial Revolution in England to really take off.

The power source for steam engines was, of course, the burning of combustible materials (wood, coal, etc.). In France in 1777, Antoine Lavoisier contemplated the basic nature of combustion.

Figure 2.17. Antoine Lavoisier.

When matter (such as wood) is burned, the amount of solid matter that remains after the burning is much smaller than the amount of material that was originally ignited. The simple conclusion from this observation was that combustion involves the conversion of matter into energy (heat). However, via a series of very clever and tightly-controlled experiments, Lavoisier was able to show that this conclusion is incorrect.

Figure 2.18. A modern reconstruction of Lavoisier's laboratory, at the Musée des Arts et Métiers, Paris.

By collecting all of the products of combustion, Lavoisier was able to show that combustion did *not* reduce the total amount of matter in the system – in fact, he was able to show that the total amount of input matter and the total amount of output matter were identical. Lavoisier was able to experimentally define a conservation law – that during a reaction, matter is neither created nor destroyed, but is conserved (even though its form may change, such as solid matter being converted into gaseous matter). This was a rather remarkable achievement, and among other things it laid the ground work for both thermodynamics and modern chemistry.

One outcome of Lavoisier's work is that it finally did away with an idea of about a century earlier. If had been postulated that the heat released during combustion is some sort of substance – and that this "substance" can also be put into some material to make it combustible. This postulated substance was known as *phlogiston* – and Lavoisier's work (by accounting for all matter both before and after the combustion reaction) showed that this "substance" did not in fact exist. This, of course, reopened the simple question of what the heat released during combustion (or, more broadly, heat in general) actually is.

The Enlightenment – What is Electricity?

While most of the scientific developments of the seventeenth and eighteenth centuries directly involved matter, heat, and light, during the eighteenth century there were also some important developments regarding electricity – which were both fundamental *and* critical to the Seebeck story.

As was noted earlier, during the first century A.D., the Roman thinker Pliny the Elder had made some simple observations regarding both the generation of – and the behavior of – what today we would call "static electricity." As electricity is a more elusive phenomenon than those more generally associated with matter, heat, and light, very little happened with electricity – until the middle of the eighteenth century. Among other things, it was inevitable that various individuals would, just by observation, note how various phenomena are similar – and that those phenomena cover a surprising range of scale.

One of those individuals was Benjamin Franklin – a notable polymath and arguably the first real American scientist.

Figure 2.19. Benjamin Franklin.

Franklin apparently happened to notice that lightning is very similar to the spark that can be induced by reaching for a metal object such as a door knob – with the only difference between the two phenomena being the differing scale (one being very large and one being very small). Franklin looked at the difference of scale and pondered a simple question: Could these be the same phenomenon at their core, with the only difference being that of a difference of scale?

This, of course, led Franklin to perform his famous experiment – in which, in 1752, he boldly (and foolishly) flew a kite during a thunderstorm, with a metal key tied to the bottom of the kite string.

Figure 2.20. Benjamin Franklin's 1752 kite-flying experiment.

As the storm drew near and lightning flashes came closer, Franklin found that by placing his finger near the key, he was able to produce a spark identical to the one that is produced by reaching for a metal doorknob. Thus, Franklin had shown that the two phenomena are indeed the same – they are both due to electricity, and are differentiated only by the vast difference in scale between them. In fact, Franklin had inadvertently demonstrated what would become one of the key (pun intended) technologically-important features of electricity – that it operates (and can be controlled) over a very large range of scale.

Franklin was lucky to survive his experiment – one that he may have been loath to even try had he truly understood both the experiment and the potential lethality of lightning. The following year (1753), a Swedish scientist, Georg Wilhelm Richmann, attempted to repeat Franklin's experiment in Saint Petersburg, Russia – and was killed by the lightning strike that he inadvertently attracted.

As Richmann's disastrous result showed, Franklin's experiment touched off intense interest in the study of electricity and electrical phenomena. And while Franklin had demonstrated that lightning was an electrical phenomenon, lightning was obviously impractical for either laboratory study or laboratory usage. However, as the notion of the vast range of scale of the same electrical phenomena had been established, other avenues could be opened.

This led to the development of very simple devices that could be used to generate electricity at-will in a laboratory setting. It was already known (by empirical observation) that electricity could be produced by spinning a coil of wire between the poles of a fixed magnet. This led to the development of hand-cranked contraptions, which were used for electrical experimentation.

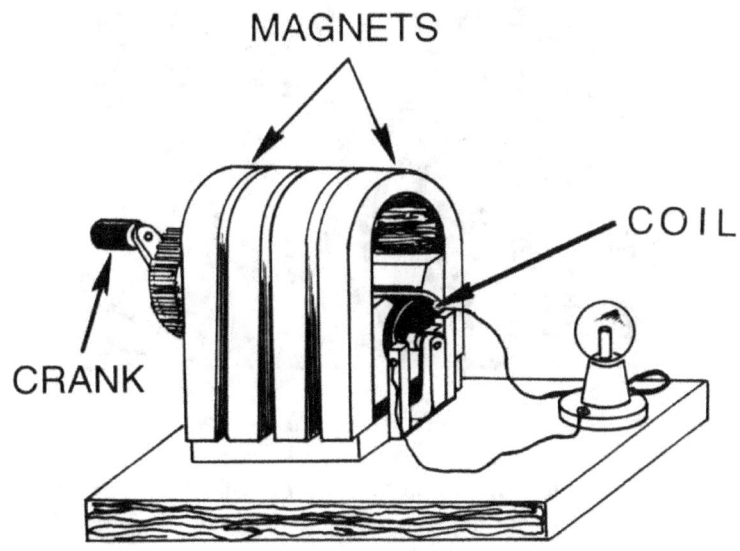

Figure 2.21. A hand-cranked laboratory apparatus for generating electricity.

The positive attributes of such devices mainly involve control; rather than having to wait for lightning to strike (literally), laboratory scientists could create their own electricity by turning the cranks. On the other hand, the production of electricity by cranking away on such a machine was fatiguing and could not be sustained for lengthy periods of time; for the same reasons, the electrical output produced by such methods was not consistent and could not be subjected to fine control.

A major improvement in electricity-generation came in 1800 – from Alessandro Volta.

Figure 2.22. Alessandro Volta.

Volta assembled a stack of alternating wafers of silver and copper, with each of those metal wafers being separated from its neighbors by a wafer of cardboard that had been soaked in brine (very salty water, which served as an electrolyte). Later, Volta found that he could replace the scarce silver with wafers of zinc – while the basic structure remained the same.

Figure 2.23. Volta's original apparatus.

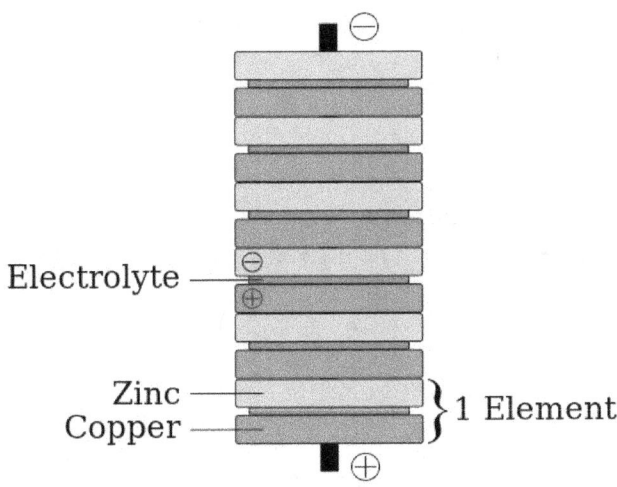

Figure 2.24. Volta's revised apparatus, using zinc disks rather than silver disks.

This device is able to produce a constant voltage – and by changing the size and scope of the stack, this constant voltage can be chosen. At the time, this device was called the "Voltaic pile;" however, it quickly took on the name by which we know it today – the "battery."

Volta's battery was epochal for a number of reasons. Most importantly for the time of its invention, the battery provided a constant and reliable source of electricity – something that was in sharp contrast to the hand-cranked generators that had been available (with the shortcomings that were described earlier). In addition, rather than design the battery stack to achieve a particular outcome (such as a specific voltage), several existing batteries could be yoked together to achieve various particular outcomes – for example, batteries could be connected in series to additively-increase the voltage, or connected in parallel to increase the current-driving ability. Of course, as is most-commonly the case today, batteries could also be used to store energy; they could be "charged up" via some input method (such as by using one of those hand-cranked generators), and then that energy could be discharged back out at will.

Ironically, Volta's battery was such a breakthrough that it has proven basically impossible (over more than two centuries) to improve it much further – today's batteries are essentially identical to the "pile" that Volta developed. This is not because people are lazy or stupid; battery technology simply offers no pathway for continuous and ongoing improvement, and no revolutionary breakthrough has appeared on the scene since Volta's time.

The Stage is Set

Thus was the scientific stage set; this is the scientific world into which Thomas Seebeck entered when he arrived in Germany in 1788.

Chapter 3: The Scientific Work

As described in Chapter 1, Thomas Seebeck had the good fortune to arrive in Germany at a very opportune time. Many new lines of inquiry into basic science were in progress or were starting up – in many places and into many topics. Seebeck was very talented, and also had a wide range of interests – as the spectrum of his work was to show.

Seebeck's most important work is surveyed in this chapter. The discussion here will not proceed chronologically. As also noted in both the Preface and in Chapter 1, Seebeck is best-known for his work on thermoelectricity – work that he actually carried out rather late in his career. Thus, Seebeck's work on thermoelectricity will be considered first – followed by other work that is equally-important but (at least to date) lesser-known.

Thermoelectricity (1821)

As noted above, among the significant scientific work carried out by Thomas Seebeck, his 1821 studies of thermoelectricity occurred later in his career; however, as it is this work for which Seebeck is best-known – and which has actually had the most visible long-term impact – it will be considered first here.

At the time that Seebeck carried out this work, the nature of both "electricity" and "heat" were still rather poorly understood. While early technological uses of electricity (such as Faraday's first motor) were being developed at this time, that Seebeck was able to do such useful work in the field is rather remarkable.

The Periodic Table of the Elements

By 1821, there were already the beginnings of some organizational understanding of the various material elements found in nature. Even though it would be several more decades (1869) before Dmitri Mendeleev introduced his concept of the periodic table of the elements, it is useful to consult such a table for evaluating the characteristics of various materials.

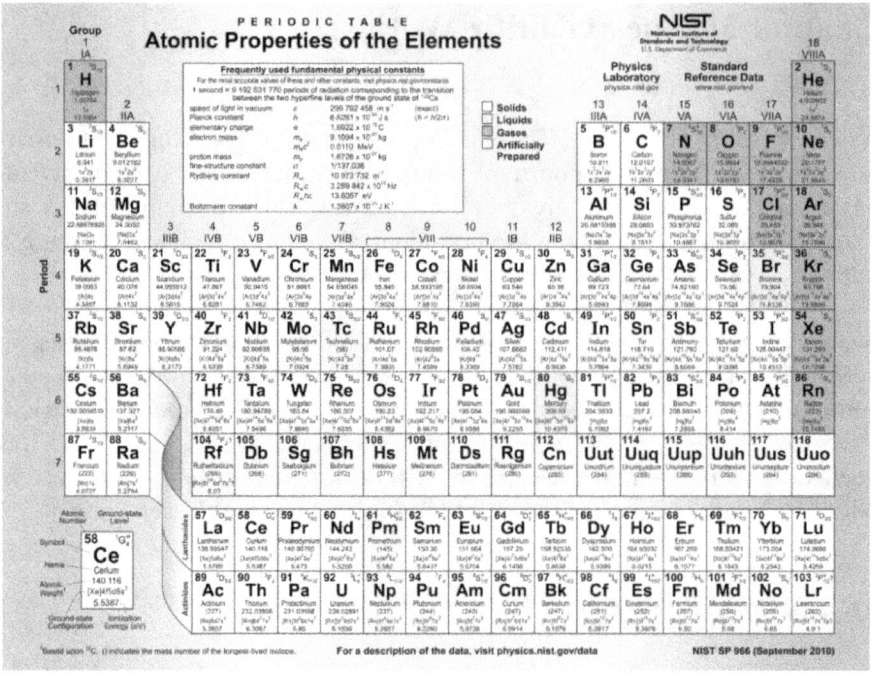

Figure 3.1. The Periodic Table of the Elements.

Some of these elements are well-known, while others are less well-known. At room temperature, most are solids – while several are gasses, and a few are liquids. Notably, the vast majority of these elements are classified as "metals." While "metals" are defined via several characteristics (that they are malleable, ductile, and possessed of "luster"), it was already known by 1821 that most metals are excellent conductors of both electricity and heat.

Naturally, both electrical conductivity and thermal conductivity are material properties that can be quantified and compared; beyond the qualitative division between "metals" and "insulators," the various elements show varying degrees of electrical conductivity (Table 3.1).

Among the elements, silver is the most electrically conductive – that is, it has the highest natural electrical conductivity. In an ideal world, silver would actually be the best material for electrical wiring; however, as is commonly known, silver is a rare precious metal – in short supply, and thus only available at relatively high cost. Copper is slightly less conductive, but is much more common than is silver – and it is thus much less expensive; this is why copper immediately became the material of choice for electrical wiring and other electrical applications.

Element	Conductivity (10^6/cm-ohm)
Silver	0.63
Copper	0.596
Gold	0.452
Aluminum	0.377
Tin	0.0917
Arsenic	0.0345
Antimony	0.0288
Bismuth	0.00867
Germanium	$1.45*10^{-8}$
Silicon	$2.52*10^{-12}$

Table 3.1. The electrical conductivity of several elements; the sidebar on the left indicates "metals" (top), "semi-metals" (middle), and "insulators" (in this case, semiconductors - bottom).

The table above shows how the electrical conductivities of different elements compare with each other; the conductivities are neither sharp nor clustered, but instead represent a continuum among these elements – with each element possessing a unique and quantifiable electrical conductivity as a fundamental material property.

Further insight can be provided – in particular – by comparing the electrical conductivity of aluminum with the electrical conductivity of tin. Aluminum has a conductivity that is about four times that of tin. While tin is commonly regarded as being a metal, tin's conductivity is actually (relatively) quite poor when compared to other metals. Thus, as the table shows, tin (along with several other elements) is classified as being a "semi-metal;" this is a sub-category, indicating that while these metals may possess the non-electrical characteristics of metals described earlier, their electrical conductivity is noteworthy for its inferiority in comparison to the other metals.

A similar table can be constructed for the thermal conductivities of the same set of material elements.

Element	Conductivity (W/m·K)
Silver	429
Copper	401
Gold	318
Aluminum	237
Tin	66.8
Arsenic	50.2
Antimony	24.4
Bismuth	7.97
Germanium	60.2
Silicon	149

Table 3.2. The thermal conductivity of several elements.

The results follow the general idea of early observations of the properties of metals – the electrical conductivity and the thermal conductivity correlate very strongly. Just as the electrical conductivity of metals is much better than that of semi-metals, so also is the thermal conductivity of metals much better than that of semi-metals. The only oddity among this set of material elements is the seemingly-high thermal conductivity (in light of the poor electrical conductivity) of the non-metallic elements germanium and silicon; this is but one indicator of some special attributes of these two elements, as will be discussed in Chapter 4.

Thermoelectricity and Electrical Conduction in Metals

In 1820, the Danish scientist Hans Christian Ørsted found (somewhat accidentally) that electricity and magnetism are related.

Ørsted built a simple circuit, composed of a single battery and a single wire – and happened to have a compass sitting close to a part of the wire. When the battery was turned on and current flowed in the wire, the compass – no matter what direction it had been pointing at the outset – would turn toward the wire.

Figure 3.2. Hans Christian Ørsted.

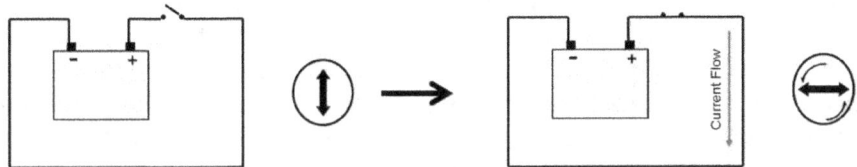

Figure 3.3. Ørsted's experiment, in which current flowing in a wire was found to attract the needle of a nearby compass.

When the direction of the current in the wire was reversed, the other end of the compass needle would point toward the wire. (A unit of magnetic field strength is known as an "Oersted" in his honor.) This was one of the earliest experiments that showed that there is some fundamental relationship between electricity and magnetism – though it would be several more decades before experiment and theory provided enough information for that unified relationship to be properly understood; this occurred in 1864, when James Clerk Maxwell was able to provide his famous set of equations describing unified electromagnetism.

In 1821, Thomas Seebeck performed a relatively simple experiment, and produced a rather remarkable result. Seebeck connected two pieces of dissimilar metals together in a loop with two junctions – and then applied heat to one of the junctions; with the heat applied, a nearby compass needle was deflected.

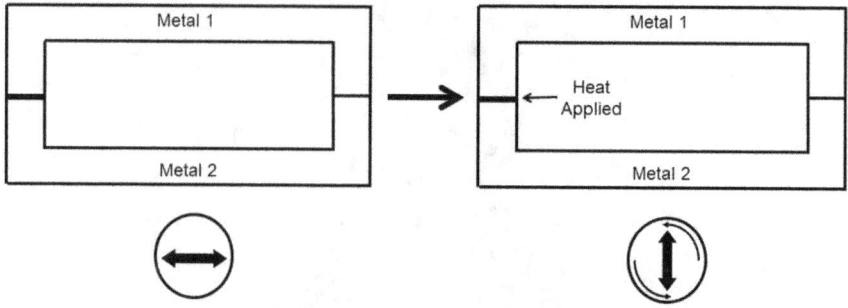

Figure 3.4. Seebeck's experiment, demonstrating how the application of heat to one junction in a loop of two dissimilar metals causes the deflection of a compass needle (thermomagnetism).

Reporting on these experimental observations in Berlin, Seebeck dubbed what he had observed as "thermomagnetism."

Upon reviewing the results, it was (not surprisingly, given his experiments of the previous year) Ørsted who was able to deduce what was going on. Since the compass needle was being deflected, this implied that (as had been done by intent in Ørsted's 1820 experiments) a current was flowing in the metallic apparatus. In addition, the heating at one end of the apparatus (but not at the other) amounted to the creation of some sort of temperature gradient along the length of that metallic apparatus. Since current was flowing (at least temporarily) in what was an open circuit, Ørsted deduced that the heat applied at one end of the apparatus was driving charge from the warm end to the cold end – and that the driven charge, being a current, was causing the compass needle to turn. Furthermore, the charge being driven to the cold end of the apparatus should be measureable there as a voltage between the ends of the two pieces of dissimilar metals.

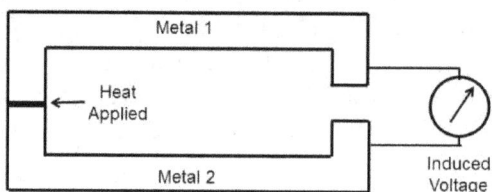

Figure 3.5. Seebeck's experiment as interpreted by Ørsted, in which the applied heat (and resulting temperature difference) induces a voltage (thermoelectricity).

Based on this evaluation, Ørsted coined the term "thermoelectricity" for what Seebeck had originally described as "thermomagnetism."

Further work showed that the relationship between the temperature gradient and the induced voltage was not at all haphazard; this relationship is linear (the induced voltage is proportional to the temperature difference between the two ends of the metal), and is quantifiable for each particular type of metal element. Mathematically, it was found that:

$$V = a \cdot (T_{hot} - T_{cold}),$$

where T_{hot} is the temperature at the hot end of the metal, T_{cold} is the temperature at the cold end of the metal, V is the voltage induced by the temperature difference, and a is what is known as the Seebeck coefficient. It was found that a is a material property that is uniquely-identifiable for any particular material. In other words, for any particular metal, the induced voltage will be proportional to the absolute temperature difference across that metal sample; that constant (a) is independent of both the voltage and the temperature(s).

This is an interesting finding; however, at the time, ideas about "charge" were only loosely-known, so there was no clear explanation of what was really going on. Somehow, temperature was causing "charge" to move in the metal – setting up the induced voltage.

One year later (in 1822), Joseph Fourier produced an equation describing the relationship between the temperature gradient across a sample of material and the heat energy flow that is thus induced.

Figure 3.6. Joseph Fourier.

Fourier found that

$$Q = -\kappa \cdot \nabla T,$$

where ∇T is the temperature gradient across the material (or, more simply and generally, the temperature difference between the two ends of the material), Q is the flow of heat energy through the material, and κ is the thermal conductivity of the material. The thermal conductivity (κ) is basically a proportionality constant for this relationship between the temperature gradient/difference and the induced heat flow; the thermal conductivity is also a fundamental physical characteristic of any material. Fourier had deduced a physical description of this heat flow behavior, but had offered no insight into the specific (micro-level) mechanisms that were causing it to appear.

In 1834, Jean Peltier conducted an experiment that was similar to Seebeck's earlier experiment.

Figure 3.7. Jean Peltier.

Using a two-metal arrangement that was configured slightly-differently than that used in Seebeck's experiments, Peltier forced a current around a two-metal ring – and found that a temperature difference was produced.

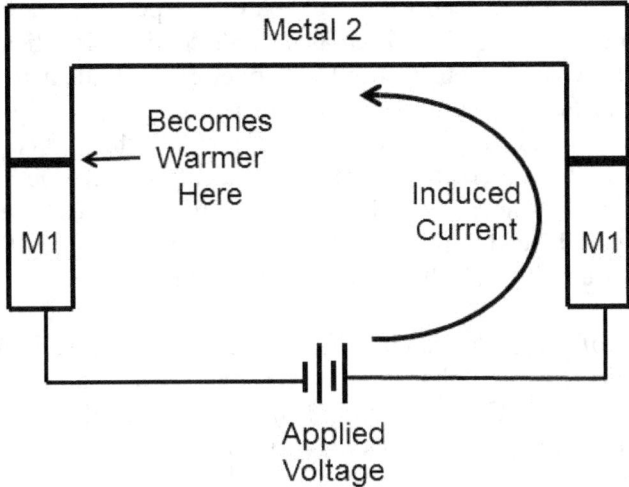

Figure 3.8. Peltier's experiment, demonstrating how a current can induce a temperature difference.

Peltier's experiment differs from Seebeck's largely in that the latter is static while the former is dynamic – that is, Seebeck's experiment involved an induced voltage but not necessarily a circuit, while Peltier's experiment involved a circuit in which current was actively flowing. This implied that the current flow was inducing a thermal flow (flow of heat) in Peltier's circuit.

In a finding similar to Seebeck's, Peltier was able to establish a simple quantitative description of the relationship between the current flow and the heat flow:

$$j_{thermal} = \Pi \cdot j_{electrical},$$

where $j_{electrical}$ is the electric current being forced through the circuit, $j_{thermal}$ is the thermal current that is being induced, and Π is what is known as the Peltier coefficient. Similar to Seebeck's finding with a, Peltier found that Π has a uniquely-quantifiable value for any particular material – as a basic material property. As with Seebeck's experiments, this was obviously a very insightful experiment that was indicating some fundamental information regarding the nature of matter – though, once again, at the time, there was insufficient theoretical knowledge to provide an understanding of either the experiment or its significance.

As might be expected, Seebeck's and Peltier's experiments are so similar that they almost beg to be identified as being essentially identical with only cosmetic differences; Seeback used a temperature gradient to produce a voltage, while Peltier used a current to produce a thermal gradient. Of course, as was noted earlier, in the early decades of the 19th century, there was only a vague understanding of the concepts of "charge," "voltage," "current," etc. – and it would only be many decades later that the "electron" would be definitively identified (Thomson, late 19th century) and properly quantified (Millikan, early 20th century).

At the time of Seebeck's and Peltier's thermoelectric experiments, the relationship between "voltage" and "current" was not even properly understood. In 1827, Georg Ohm published the results of his studies of the behavior of the current and voltage in various metals.

Figure 3.9. Georg Ohm.

As his now-famous name implies, it was Ohm who first postulated what came to be known as "Ohm's Law" regarding the relationship between voltage and current:

$V = I \cdot R,$

where V is the applied voltage, I is the current produced by the applied voltage, and R is the proportionality constant that Ohm found to explain the behavior – which is, of course, the "resistance" of the material in question.

Ohm's findings easily unify Seebeck's and Peltier's experiments; as Peltier's circuit has a current flowing, there has to be a voltage along the circuit in order for this to happen; Seebeck's experiment differs only in that it is an open circuit where no DC current can flow.

In his 1827 work, Ohm also provided a more-generalized form of Ohm's Law as written above. Ohm was also able to show that:

$$J = \sigma \cdot E,$$

where E is the electric field across the material, J is the current density (the current per unit cross-sectional area in the direction of the current flow) induced by the applied field, and σ is the electrical conductivity (such as those tabulated in Table 4.1). Once again, this is a fundamental finding – but one where the actual underlying explanation was not readily available at the time.

Of further note, following his initial 1821 work, Seebeck repeated his experimental studies using a wide variety of materials – including both metals and semiconductors. He noted in particular that when the two materials in the experiment involved a combination of a metal and a semiconductor (rather than another metal), semiconductor materials showed much stronger thermoelectric behavior than did metals; this was an early indication that semiconductors are somehow fundamentally different from metals (and insulators). Further, Seebeck noted that bismuth exhibited some very strange behavior (which will be discussed in Chapter 4) – and that other materials also exhibited this "bismuth-like" behavior.

Light and Matter – Pondering Color, 1810

Since time immemorial, humans have been fascinated by light and color; however, for centuries and millennia, light and color were more the province of artists than of scientists. Scientific work was scanty at best; in 1665, Isaac Newton first demonstrated that ambient light could, using a prism, be separated into its constituent components of different colors.

Newton's studies of light and color continued for the rest of his life – resulting most notably in the publication of his *Opticks* in 1704.

But when it came to the idea of "capturing" light and color in a recordable, essentially-permanent form, it was still the artists who had an exclusive ability to do so – via painting. But painting itself almost seemed to ask the question – was there some way to capture light directly, and thus to create an image as-is?

Prior to the nineteenth century, work in this area had been rather limited. In 1727, Johann Heinrich Schultze made an interesting accidental discovery. Working in his laboratory one day, he happened to leave a small lab dish of material – a powder that was a combination of silver nitrate and chalk – on a table; as luck would have it, a shadow line fell across the middle of the dish of powder on the table. Returning later to that dish on the table, Schultze saw a "shadow line" in the powder that at first appeared to be simply sunlight on one part of the dish and shadow on the other part of the dish. However, when he picked up the dish to move it, he was startled to see that the "shadow" had been taken up by the powder – the part that had been exposed to the sunlight was lighter than the part that had remained in the shadow. Schultze had inadvertently found a material that would become lighter when exposed to light but remain dark when not exposed to light.

In 1810, Thomas Seebeck found something even more remarkable. While Schultze's powder turned lighter when exposed to light, Seebeck found that chlorides of silver are able to take on the color of the light to which they are exposed.

Further, Seebeck noted that there was apparent "light" beyond violet – "light" which was invisible to the human eye but which was clearly causing his experimental material to change color. This was an early observation of what eventually came to be known as "ultraviolet light" – "light" invisible to the human eye, in wavelengths just beyond the wavelength of violet light.

At this time, Seebeck also worked closely on various scientific aspects of light with Johann Wolfgang von Goethe. This is, perhaps surprisingly *the* Goethe of literary fame – but like Seebeck, Goethe was also a polymath of numerous varied interests and talents. In fact, in the same year that Seebeck carried out his aforementioned experiments with chlorides of silver, Goethe published a treatise entitled "Theory of Colors" – concerning the perception of color by humans.

Another contribution made by Seebeck in this area was his work in what is known as photoelasticity or piezo-optics. Credit for the first work in this area is usual given to the Scottish physicist David Brewster, due to his publications (in English) in 1815 and 1816; however, Thomas Seebeck published his first work in this area (in German) in 1813 and 1814.

Photoelasticity (or piezo-optics) refers to a phenomenon that was first observed by Seebeck – that stress in transparent materials will alter their optical properties. In particular, stress in a material will cause what is known as *birefringence*. Birefringence is a phenomenon that had previously been observed in some transparent crystalline materials; these materials actually contain regions with different refractive indices – and thus, light passed through them will separate (to some degree) and produce colors in various patterns.

It was immediately obvious that this birefringence phenomenon could be used to examine stress (under external forces) in materials with any degree of pliability.

Figure 3.10. Birefringence patterns induced by stress in a piece of transparent plastic.

Patterns of this type will change when the stress on the material is changed (in either direction or strength).

Due to their work with this phenomenon, Seebeck and Brewster shared an 1815 prize that was given by the French Institute for the best research in physics over the preceding two years.

Early Biochemistry – The Optical Activity of Sugar (1818)

Another important contribution made by Thomas Seebeck provided the initial foundations of what was eventually to become biochemistry.

Once the ability to filter light – so as to produce an output of polarized light – had been created, it provided another tool for scientific investigations of various sorts. Seebeck (basically simultaneously with the French scientist Jean-Baptiste Biot), noted that when polarized light is passed through a solution of sugar dissolved in water, the plane-of-polarization of the light is rotated:

> Biot and Seebeck pointed out in 1815 that certain organic substances have the power to rotate the plane of polarization. Oil of turpentine, and sugar and tartaric acid in aqueous solution, have this property, as was shown at this early date.

(From H. C. Jones, "The Elements of Physical Chemistry," 1902.)

These early observations were, of course, something of a mystery. However, it was clear that complex carbon-based substances that were involved in biological systems were able to show a number of unique properties that were not observed with simpler materials.

Chapter 4: The Legacy

As was noted earlier, Thomas Seebeck is best-remembered for his work on thermoelectricity and thermoelectric behavior; however, he also made important contributions in other fields.

Seebeck's work has had a lasting impact in all of those fields – and that is still the case today.

Thermoelectricity

Seebeck's 1821 experiments in thermoelectricity combined together two phenomena that, at the time, were just beginning to receive serious scientific study – heat and electricity. In the immediate aftermath of Seebeck's work, the study of these two phenomena diverged for several decades.

The study of heat developed into the study of thermodynamics – with much of the concentration (both theoretically and practically) being connected to the behavior and operation of steam engines, as the steam engine was the most important technology of the mid-nineteenth century. The study of electricity developed into the study of electricity and magnetism in tandem – which led eventually to the development of large-scale electromechanical technology; this eventually led to early efforts in power generation and transmission, along with early developments in communications technology.

It was several decades after Seebeck's 1821 work that these two fields began to have significant overlap once again.

Thermodynamics

In 1824, Sadi Carnot, a French military engineer, published a short but remarkable treatise on the nature of heat and motion.

Figure 4.1. Sadi Carnot

RÉFLEXIONS

SUR LA

PUISSANCE MOTRICE

DU FEU

ET

SUR LES MACHINES

PROPRES A DÉVELOPPER CETTE PUISSANCE.

PAR S. CARNOT,

ANCIEN ÉLÈVE DE L'ÉCOLE POLYTECHNIQUE.

———

A PARIS,

CHEZ BACHELIER, LIBRAIRE,

QUAI DES AUGUSTINS, N°. 55.

1824.

Figure 4.2. The title page of Carnot's 1824 book.

In this book (which although largely ignored in Carnot's time, is still in print today), Carnot essentially laid out the basis for what would become thermodynamics – and even discussed the concepts that would eventually come to be known as entropy and the second law of thermodynamics.

In 1834, two years after Carnot's untimely death, his work was discovered by the French scientist Benoît Clapeyron.

Figure 4.3. Benoît Clapeyron.

Clapeyron reinterpreted much of Carnot's thoughts into a more-accessible form – including his recasting of Carnot's thoughts on the expansion and contraction of gasses into a graphical form of plotting pressure against volume, with induced changes proceeding along isothermal contours in that pressure-volume space.

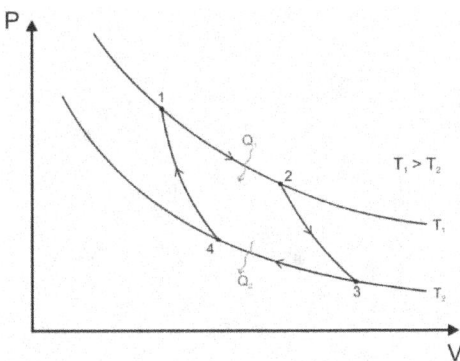

Figure 4.4. Clapeyron's interpretation of Carnot's thoughts on the pressure and volume behavior of a gas with intentionally-changed temperature conditions.

This interpretation of the cycling of pressure and volume with controlled changes in temperature is known as the Carnot cycle – and it specifically describes the ability of such piston-based systems to do work. Clapeyron understood the importance of this cycle, as it allowed for optimal use of the pistons in steam engines (as Clapeyron wrote in 1842). In addition, in 1843, Clapeyron further developed Carnot's ideas about reversible and irreversible processes – which eventually came to be known as the second law of thermodynamics.

In 1845 in England, James Joule published remarkable work on the relationship between heat and work.

Figure 4.5. James Joule.

Joule used a simple apparatus to study the heating and cooling of a confined liquid when work was done *to* that liquid and *by* that liquid.

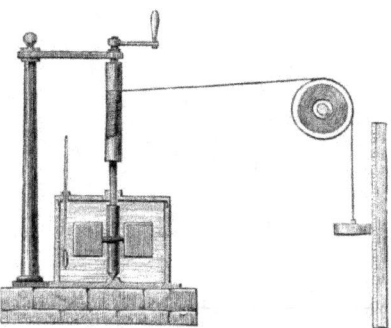

Figure 4.6. Joule's apparatus for studying the relationship between work and heat; the falling weight (work) turns the shaft, which agitates the water; this motion creates heat, which raises the temperature of the water.

Critical to Joule's work was the availability to him of very accurate thermometers – which allowed him to measure the temperature (and changes in temperature) in the liquid in the apparatus; such very accurate thermometers were available to Joule not because he was a scientist, but because he was a brewer. Joule's key finding was that "heat" and "work" are interchangeable – heat can produce work, and work can produce heat.

In 1847, Hermann von Helmholtz gathered together the accumulated findings.

Figure 4.7. Hermann von Helmholtz.

Helmholtz formalized the earlier observations with his own overarching conclusion – that just as Lavoisier had earlier observed that matter can neither be created nor destroyed (but only converted from one form to another), in the same way energy can neither be created nor destroyed, but also only converted from one form to another. This is the first law of thermodynamics.

In 1851, Lord Kelvin was able to confirm what Francis Bacon – some 250 years earlier – had surmised.

Figure 4.8. Lord Kelvin.

Heat is not a substance, but is the consequence of mechanical effects – that is, as Bacon had earlier inferred, heat is indeed a manifestation of motion of some sort (or sorts).

In the context of Seebeck's 1821 work on thermoelectricity, this was an interesting and unifying finding. During the intervening thirty years, "heat" and "electricity" followed separate scientific paths; as described above, the "heat" path led to the formulation of the laws of thermodynamics (which found practical application in steam engines), while the "electricity" path led to the gradual accumulation of knowledge about both electricity and magnetism (which found practical application in large electromechanical equipment, such as generators, motors, and lighting).

Kelvin's conclusions brought these two branches back into contact. In terms of Seebeck's (and Peltier's) work on thermoelectricity, it was now clear that the electrical behavior associated with Seebeck's voltages and Peltier's current had to be a manifestation of the motion of something. But what exactly was that "something"?

The Electrical and Thermal Behavior of Solids

In 1853, Gustav Wiedemann and Rudolph Franz observed that for a particular given temperature, the ratio of the thermal conductivity to the electrical conductivity was pretty much the same for a very large number of metals.

The basis of this observation is made readily-apparent by plotting the electrical conductivity against the thermal conductivity, using the four metals and four semi-metals that were presented in Table 3.1 and Table 3.2.

Given Seebeck's earlier observations, this is an important additional observation – as it clearly indicates not only that the thermal conductivity and the electrical conductivity are closely related, but more specifically that thermal conduction in metals is being facilitated by the electrical charges in that metal.

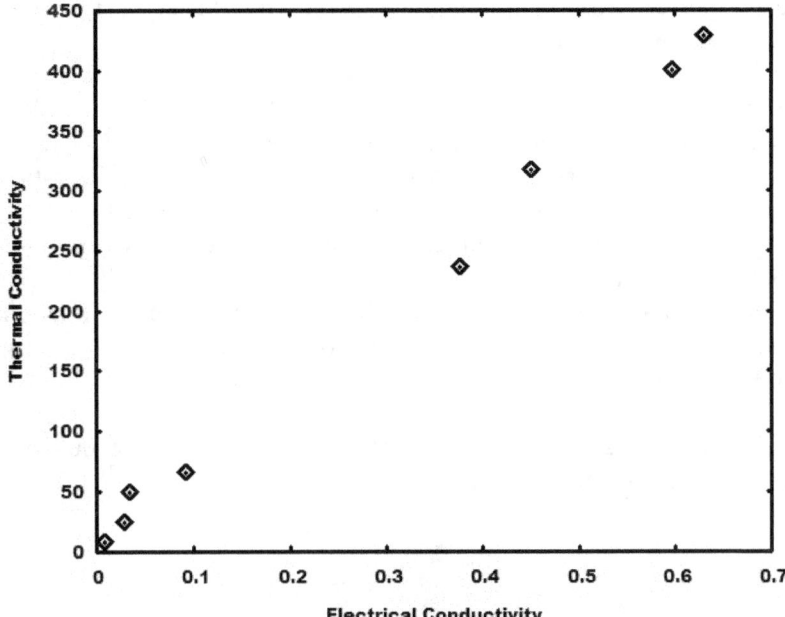

Figure 4.9. A plot of the electrical conductivity vs. the thermal conductivity for four metals (silver, copper, gold, and aluminum) and four semi-metals (tin, arsenic, antimony, and bismuth).

This observation was extended by Ludwig Lorenz, who in 1872 noted that the ratio of the two conductivities was in fact proportional to the absolute temperature; that is, for some particular value of T,

$$\frac{\kappa}{\sigma} \propto T,$$

where κ is the thermal conductivity, σ is the electrical conductivity, and T is the absolute (Kelvin) temperature. From this, Lorenz was able to find an absolute number that collected all of these terms together:

$$\frac{\kappa}{\sigma \cdot T} = L,$$

where L is what is known as the Lorenz number – which is a fundamental constant of a sort that is independent of both the specific conductivity properties of any particular metal and the particular temperature. From simple theory, Lorenz calculated a value for L of 2.45×10^{-8} W-Ω/K^2.

While not perfect, for most metals the results did cluster fairly closely to the Lorenz number L.

These results once again clearly indicated that there is a close relationship between the thermal conductivity and the electrical conductivity – and that as per Kelvin's observations (that heat is indeed a manifestation of motion), the thermal conductivity behavior of a metal is due to the motion of the electrical charge in that metal. However, as there was at the time no reasonable theoretical structure for the behavior of the electrical conductivity, there was no route available for linking these two phenomena (thermal conductivity and electrical conductivity) together in a more concrete fashion.

Another fundamental and interesting experiment was carried out in 1879 by the American scientist Edwin Hall.

Figure 4.10. Edwin Hall.

Hall began his experiment by simply using a voltage to force a current along the long axis of a piece of metal. This idea itself wasn't new, of course, but then Hall added to his experiment – by including a DC magnetic field, applied up through the narrow axis of the metal. Hall found that when he ran this experiment, he was able to induce a voltage (electric field) along the axis perpendicular to the axis of the current flow – that is, along the third axis of the metal slab under investigation.

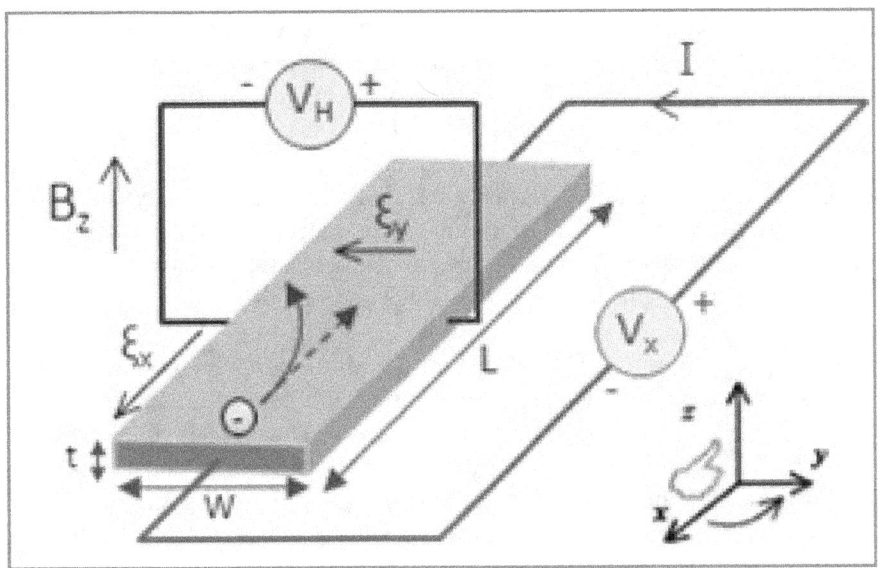

Figure 4.11. Hall's experiment, in which the applied magnetic field (B_z) induces a lateral electric field (E_y).

The induced lateral voltage V_H came to be known as the Hall voltage. Further, as was the case with Seebeck and Peltier, Hall was able to describe his results in a form where a unique coefficient appears:

$$R_H = \frac{E_y}{j_x \cdot B_z},$$

where R_H is the "Hall coefficient," E_y is the induced lateral electric field, j_x is the current density (current per unit cross-sectional area across the direction of the current flow), and B_z is the magnetic field. The Hall coefficient also turns out to be a fundamental material property, with values that correspond uniquely to each particular material.

Hall's observations amount to a very useful set of insights into the behavior of charged particles in a solid material; as there are many variables, quantitative knowledge of all-but-one of them can allow for the remaining variable to be determined. As an example, a device known as a "Hall effect magnetometer" can be used to (very accurately) measure magnetic field strength. Alternatively, the charge density in a conducting solid can be determined from Hall Effect measurements.

However, at the time that Hall conducted his experiments, a qualitative – but much more fundamental – result was produced. As was described earlier, the electric field that occurs as a result of the experimental arrangement is due to the lateral segregation of the charge that is otherwise flowing along the flowing-current axis. Hall's results on (most – but more about that later) metals demonstrated that the charged particle that is actually performing the conduction is negatively-charged – that is, conduction in metals is occurring due to a flow of electrons (rather than via a flow of positively-charged particles, such as protons).

As should be clear (and was earlier noted to be the case with steam engines), "theory" does not lead "experiment" in some fundamental way – instead, each provides ongoing fodder for the other as they advance together. A reasonable theoretical description of the observations of Seebeck, Peltier, and Hall (as well as others) did not appear until Paul Drude published his theory of electrical conduction in metals in 1900.

Figure 4.12. Paul Drude.

The main basis for Drude's theoretical analysis was to build on an observation that went back to (at least) Ohm's work in 1827 – particularly his observation that $J = \sigma \cdot E$. This observation implies that the conduction of electrons through a metal behaves in the same manner as does the movement of "particles" through a viscous medium (such as a gas or a liquid) – in which the background medium provides a uniform drag force against the force moving the particles forward.

Among other things, this allowed Drude to provide a unified explanation of the relationship between electrical conduction and thermal conduction in metals. By assuming that the "electron gas" inside the metal could be treated as obeying the by-then well-known laws for an ideal gas, Drude was able to find the same relationship between thermal conductivity and electrical conductivity that Lorenz had found some thirty years earlier:

$$\frac{\kappa}{\sigma \cdot T} = L.$$

However, while Lorenz had found a value for L of 2.45×10^{-8} W-Ω/K^2, Drude's result was 1.11×10^{-8} W-Ω/K^2 – of the same order of magnitude, but notably smaller.

Later analysis showed that Drude's surprisingly-good quantitative results (on a variety of fronts) were someone fortuitous – in that some of his assumptions were in fact off (quantitatively) by quite a bit, but by good fortune these erroneous assumptions actually canceled each other out; in addition, Drude made a calculation error that originally caused his computed value for L to be much closer to Lorenz's number than it should have been. But Drude's theoretical analysis did (among other things) serve to provide a more concrete and more formalized connection between a metal's thermal conductivity and its electrical conductivity.

Basically, by 1900, it seemed that the basic nature of electrical conduction and thermal conduction in metals was clear. Electrical conduction is provided by mobile, negatively-charge particles (electrons), which move (under the impetus of an electric field) through a positively-charged background that provides a frictional-drag resistance force – and where the ability of a particular metal to carry an electrical current is a fundamental-and-quantifiable material property unique to each metal, and is known as the electrical conductivity (σ).

With respect to the thermal conductivity, referring back to Seebeck's 1821 experiment, the heat applied to one end of the metal provides extra energy (and thus velocity) to the electrons in the metal; therefore, electrons at the "warm" end of the metal will preferentially move away from the heat source and become more concentrated toward the cooler end; this provides the segregation of charge that appears as a voltage. In Peltier's experiment, the same thing is happening, but somewhat in reverse – the applied voltage and forced current are pushing electrons to the other end of the metal, where the higher concentration of electrons, by their own motion, induce heat. In Hall's experiment, the magnetic field is displacing electrons laterally, whereby they accumulate along the side of the metal strip and create an electric field laterally across the metal strip; this lateral accumulation of charge – and the resulting lateral electric field – eventually (exactly) counterbalances the magnetic field's forcing of electrons in the lateral direction.

Often in physics, a robust and wide-ranging theory seems to be complete – except for a few loose ends. This proved to be the case with the early-20th century understanding of conduction in metals. In addition to the electrical conductivity σ, the experiments of Seebeck, Peltier, and Hall produced their own constants that are fundamental to each particular type of material in question. The explanation of these experiments – in terms of free electrons in metals – seemed to be complete; electrons are moved inside the material in a manner that is qualitatively consistent, and theoretical analysis provided a proper explanation for the quantitative values found for the relevant coefficients.

However… certain metals exhibit unusual properties. In some metals, the signs of the aforementioned coefficients are "backwards" – that is, the signs of the associated coefficients are the *opposite* of what the simple theory implies that they should be. For example, certain metals exhibit a Seebeck coefficient that is negative.

Element	Seebeck Coefficient (μV/K)
Selenium	900
Tellurium	500
Germanium	440
Silicon	300
Antimony	47
Molybdenum	10
Cadmium	7.5
Tungsten	7.5
Gold	6.5
Silver	6.5
Copper	6.5
Rhodium	6.0
Tantalum	4.5
Lead	4.0
Aluminum	3.5
Carbon	3.0
Mercury	0.6
Platinum	0.0

Sodium	-2.0
Potassium	-9.0
Nickel	-35
Bismuth	-72

Table 4.1. The Seebeck coefficient of several elements[1].

This is something of a catastrophe – as it indicates that for certain materials, the theoretical understandings are not failing in just some minor, quantitative way (e.g., via minor details which are quantitatively small and which are not fundamental to the underlying understanding). Instead, the theoretical understandings fail *completely and totally*. The "backwards" (i.e., negative) values for some of the coefficients (here, for the last four elements listed in Table 4.2) seem to imply that in certain metals, conduction is occurring via some form of positively-charged electrons! The theoretical explanation – based on negatively-charged electrons – is at a total loss to explain this behavior; a century of experiments had never found even the slightest indication of the existence of "positively-charged electrons." Something important was clearly missing in the understanding of conduction and conductivity.

However, there is (what was in the nineteenth century) a sidelight to the story – one that was to move from the sidelines to the center of the playing field of condensed-matter physics during the twentieth century. (And as was mentioned in Chapter 3, Thomas Seebeck had noted strange thermoelectric behavior in some of the materials that he had studied – such as metallic bismuth and various semiconductor materials.)

This sidelight can be introduced by revisiting a table of the electrical conductivities of several elemental materials that was presented in Chapter 3. Among the insulators are the elements germanium and silicon. The conductivities of germanium and silicon are very poor – not as poor as those of various other insulators, but very poor nonetheless (particularly when compared with the metals). However, the conductivities of germanium and silicon are just not quite bad enough for them to be completely grouped with the other insulators – as they possess a small amount of natural conductivity. This led to silicon and germanium (sometimes) being separately classified (by stretching the boundaries of language) with their own label – that they are "semi-conductors."

---

[1] Available at: http://www.electronics-cooling.com/2006/11/the-seebeck-coefficient/ .

Ironically, this ambiguity seemed to condemn germanium and silicon to uselessness – they were not conductive enough to serve the purposes served by metals, yet they were not (truly) insulating enough to the serve the purposes served by insulators!

Element	Conductivity (10^6/cm-ohm)
Silver	0.63
Copper	0.596
Gold	0.452
Aluminum	0.377
Tin	0.0917
Arsenic	0.0345
Antimony	0.0288
Bismuth	0.00867
Germanium	$1.45*10^{-8}$
Silicon	$2.52*10^{-12}$

Table 4.2. *The electrical conductivity of several elements; the sidebar on the left indicates (top to bottom) "metals," "semi-metals," and "insulators" (in this case, semiconductors).*

Among the insulators are the elements germanium and silicon. The conductivities of germanium and silicon are very poor – not as poor as those of various other insulators, but very poor nonetheless (particularly when compared with the metals). However, the conductivities of germanium and silicon are just not quite bad enough for them to be completely grouped with the other insulators – as they possess a small amount of natural conductivity. This led to silicon and germanium (sometimes) being separately classified (by stretching the boundaries of language) with their own label – that they are "semi-conductors." Ironically, this ambiguity seemed to condemn germanium and silicon to uselessness – they were not conductive enough to serve the purposes served by metals, yet they were not (truly) insulating enough to the serve the purposes served by insulators!

As the fundamental electrical properties of the various elements were quantified, another oddity of these semiconductor elements became apparent. The electrical conductivity of an element is a fundamental material property of that particular element – with an associated quantitative value that can be determined by measurement. Naturally, it was important that these values be determined and tabulated; of course, it is a relatively simple matter today to "go look up" the electrical conductivity of any particular element – such as in the table (Table 4.2) presented above.

The electrical conductivity of some material is actually quite simple to determine; using a voltage and a current, the resistivity of the material is easily measured – and electrical conductivity is simply the inverse of that resistivity. More importantly, for most of the elements, it was not difficult to get suitable and consistent numbers; as long as the sample was relatively "clean" (not overly-imbued with impurities of some sort), good numbers were readily obtained – and these numbers that did not change when a "cleaner" sample became available for analysis.

Germanium and silicon, however, provided a stark exception to this easy situation. It was quickly noted that the material conductivities of these semiconductor elements are maddeningly-difficult to determine. Unlike most of the other elements, the electrical conductivities of germanium and silicon were found to be extremely sensitive to trace impurities in the sample being used for the analysis. Even more infuriating, determining the "real" electrical conductivities proved to be difficult – as unlike the other elements, it was difficult to establish a "floor" of impurity concentration below which the electrical conductivity would no longer change; every time a cleaner sample became available, a lower value of the electrical conductivity was found.

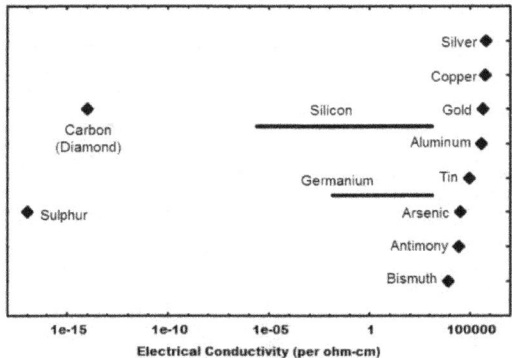

Figure 4.13. The electrical conductivity of several (elemental) insulators, semiconductors, and metals; while the conductivity of the insulators and metals are well-defined via a single, unique value for each material, the conductivity of semiconductor materials can vary over a very wide range.

Thus, while the electrical conductivities of metals and insulators are points, the electrical conductivities of semiconductors (such as germanium and silicon) show a wide spread.

Eventually, it came to be understood that this isn't a bug – but a feature (as depicted in Figure 4.13). Among metals and insulators, the electrical conductivity is an anchored, fundamental material property that simply is what it is – it cannot be adjusted or controlled. However, in semiconductors, the sensitivity of the electrical conductivity to impurities is not haphazard; for some concentration of a particular impurity in the material, a particular electrical conductivity results. In addition, as the figure above shows, the range of values over which the electrical conductivity can be varied (or *be tailored*) is extremely large – some five orders of magnitude in germanium and nearly *eight* orders of magnitude in silicon.

The engineering methods for using (and controlling) this interesting property are actually quite straightforward. Basically, a sample of a semiconductor material is placed in proximity to some "source" of impurity atoms, and these impurity atoms are allowed to diffuse into the material until they have completely and uniformly permeated the material to some level of a uniform impurity concentration. By controlling the concentration of impurities diffused into the silicon, the electrical conductivity of the silicon can be chosen – over a rather wide range of possible conductivities.

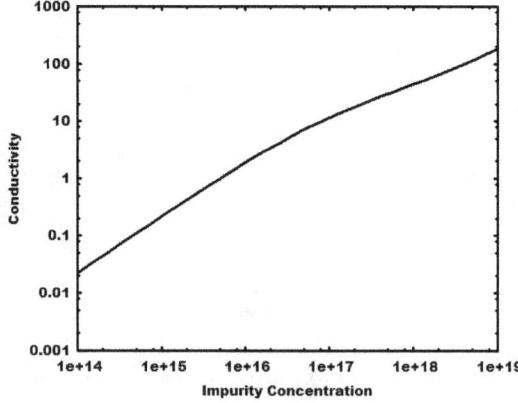

Figure 4.14. The electrical conductivity of silicon as a function of the introduced impurity concentration in the silicon.

A variety of impurities can be introduced into silicon to produce this behavior. When atomic species such as phosphorous and arsenic are diffused into silicon, they provide free electrons that allow the silicon to take on an electrical conductivity – behavior similar to that in a metal, but with a lower level of conductivity. However, when certain other atomic species are diffused into silicon, something completely different happens. When boron is diffused into silicon, the silicon will also take on a controllable level of electrical conductivity. However, in this case, something completely different – qualitatively – happens. In this situation, the conductivity of silicon is effectively backwards – that is, the conductivity behaves as if the free charge in the material (which provides the conductivity) is positive rather than negative. This, of course, reprises the behavior observed in certain metals – where the Seebeck and Peltier coefficients are "backwards," implying conductivity by positively-charged particles which (unlike negatively-charged electrons) are never observed in nature.

This odd dichotomy is best shown via an examination of the behavior of the Peltier coefficient of silicon – for two samples, each of which has had a different types of impurity diffused into it (Figure 4.15). In silicon that has received electron-giving impurities (such as phosphorus or arsenic), the behavior of the Peltier coefficient is normal; however, when silicon has been infused with an impurity such a boron, the Peltier coefficient is backwards – confirming that the conduction mechanism involves what amounts to positive charge.

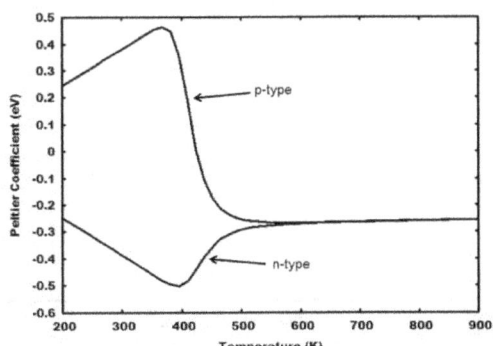

Figure 4.15. The Peltier coefficient of two samples of silicon as a function of temperature, where one sample has been infused with phosphorous ("n-type") while the other has been infused with boron ("p-type"); the terms "n-type" and "p-type" are explained below. (Data courtesy of Bart Van Zeghbroeck.)

(The same type of result is observed if the Seebeck coefficient is used – since, as described earlier, the Seebeck and Peltier methods are essentially the same thing.)

The theoretical explanations that eventually described this seemingly-odd behavior in semiconductors also served to finally close up the long-known problem of the "backwards" behavior of the observed conduction in certain metals, such as bismuth. It is indeed possible for situations to arise where the electrical conductivity does behave as if positively charged particles – rather than negatively-charged electrons – are responsible for the conduction. This is something of a ghostly situation, as these "positively-charge carrier particles" cannot be observed outside of the solid material itself!

When a semiconductor has been infused with impurities that cause it to have conductivity due to negative charge, that semiconductor is said to be *n-type* – with the "n" denoting negative-charge behavior. Conversely, when a semiconductor has been infused with impurities that cause it to have conductivity due to (apparent) positive charge, that semiconductor is said to be *p-type* – with the "p" denoting positive-charge behavior.

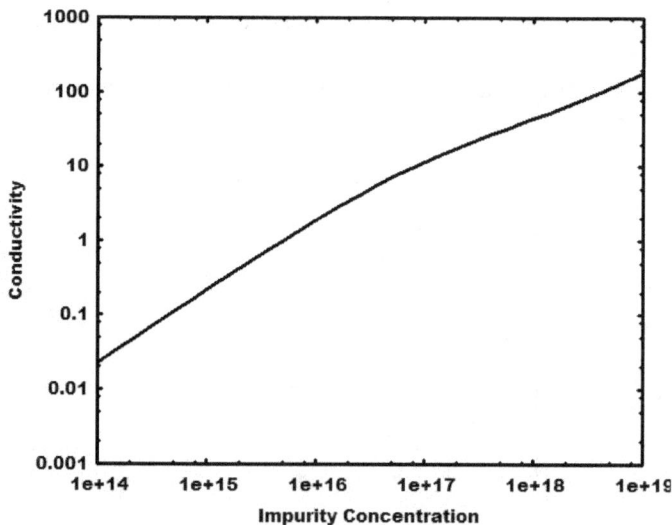

Figure 4.16a. The electrical conductivity of n-type silicon as a function of the introduced impurity concentration in the silicon.

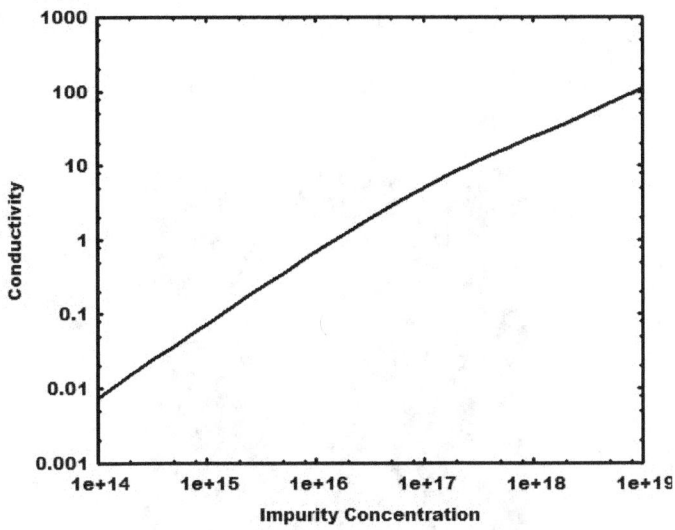

Figure 4.16b. The electrical conductivity of p-type silicon as a function of the introduced impurity concentration in the silicon.

Thus, Seebeck's 1821 experiments provided a method of material analysis that was to lead to the discovery "oddities"; these oddities ended up providing the basis for semiconductor technology – which has turned into the largest and most-important technological juggernaut in human history.

Superconductivity

However, with respect to the behavior of condensed matter, Seebeck's methods led to another important insight.

As was noted empirically and theoretically, there is a close, parallel relationship between electrical conductivity and thermal conductivity; a material that is a good electrical conductor will (generally) also be a good thermal conductor, and a material that is a poor electrical conductor will (generally) also be a poor thermal conductor. Essentially, the electrons in a solid material serve to transport entropy. If a material is more electrically conductive, the electrons will more-readily transfer thermal energy – while the opposite is true of a material which is less electrically conductive. Further, as per Seebeck's observations regarding how a larger thermal gradient across the material will produce a larger voltage, "hotter" electrons will more easily move toward the other end of the material than will "cooler" electrons.

However, in 1911, some very strange measurement results were found by Heike Kamerlingh Onnes.

Figure 4.17. Heike Kamerlingh Onnes.

Kamerlingh Onnes had earlier led a team that developed methods of liquefying various gases by finding ways to reach very low temperatures; this led to a notable first in 1908, when that team liquified helium (something which occurs at a shockingly-low temperature of 4.2K).

With the ability to reach low temperatures in the laboratory, Kamerlingh Onnes began investigating the behavior of materials at these newly-accessible temperatures. In 1911, Kamerlingh Onnes was measuring the conductivity of mercury as the temperature was lowered – and got a shock. At the temperature of liquid helium (4.2K), the electrical resistivity of mercury disappeared – and the resistivity was not "small," but actually *zero*.

Since the conductivity of mercury (and, as was later found, of other metals as well) did not simply continue to decrease linearly with temperature – but instead collapses to zero, at a particular temperature specific to that metal – it was already obvious (as Kamerlingh Onnes immediately realized) that this phenomenon was a manifestation of something that was very unusual.

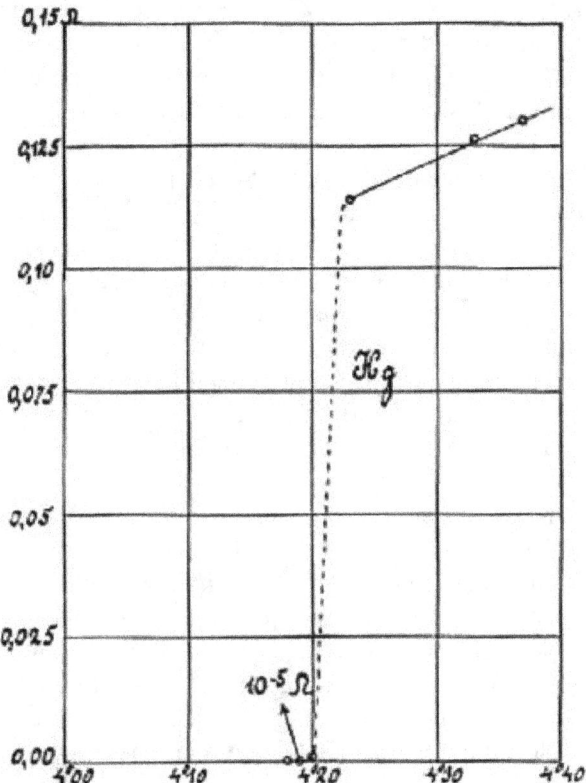

Figure 4.18. The electrical conductivity of mercury as a function of temperature.

The behavior of the electrical conductivity is indeed something completely different; this is not just "conductivity," but is what came to be known as *superconductivity*. Not surprisingly, it took decades for solid theoretical explanations of superconductivity to be developed.

However, the methods developed by Seebeck were quickly able to provide a stark indication of how truly different superconductivity is from all prior known material behavior. As noted above, following on Seebeck's 1821 observations regarding thermoelectricity, further succeeding developments clearly indicated a very close relationship between electrical conductivity and thermal conductivity.

A superconductor is not merely a good conductor; it is a *perfect* conductor, with an electrical resistance that is truly *zero*. However, a superconductor is actually a *very poor thermal conductor*; a superconductor exhibits no Seebeck effect (and no Peltier effect). This is a direct contradiction of the earlier results on metals (and other materials) – where the electrical and thermal behavior are closely-complementary of each other.

Thus, the insights developed from Seebeck's observations quickly provided confirmation that superconductivity is indeed something very different from all prior knowledge in solid state physics. Theory had cleared demonstrated that the good thermal conductivity observed in metals was directly due to the good electrical conductivity – as electrons transport entropy (and thus heat) through the metal. Therefore, even without any theoretical explanations of superconductivity, it was clear that when a metal is in a superconductive state, *the electrons transport no entropy*. This is a rather astonishing observation.

Eventually, theoretical explanations of superconductivity were developed – and they are indeed rather stunning, as detailed quantum mechanics are required to explain what is occurring. Electrons are able to pair up into a single quantum state – and thus they do not behave as do electrons in metals at more-normal temperatures. Electron flow in a normal metal is "disorderly" and thus induces an accompanying thermal current; in contrast, electron flow in a superconductor is orderly – and thus is non-entropic, and so does not create an accompanying thermal current.

The work done by Seebeck nearly a century before the discovery of superconductivity (and nearly a century and a half before a suitable theoretical explanation of superconductivity was achieved) allowed for critical insights into the special nature of superconductivity.

The "Peltier Cooler"

As was described in Chapter 3, it quickly became obvious that thermoelectric behavior is a fundamental property of any material – and that the temperature-dependent behavior of a material followed a rather simple law over a rather large range of temperatures:

$$V = a \cdot (T_{hot} - T_{cold}),$$

where T_{hot} is the temperature at the hot end of the metal, T_{cold} is the temperature at the cold end of the metal, V is the voltage induced by the temperature difference, and a is what is known as the Seebeck coefficient.

As also described in Chapter 3, *a* is a material property that is uniquely-identifiable for any particular material.

This simple material property is quite useful – as it can be exploited to create thermometers that are both extremely accurate *and* which retain that accuracy over large temperature ranges.

However, one of the more practical applications of Seebeck's work in thermoelectricity is what is known as a "Peltier cooler."

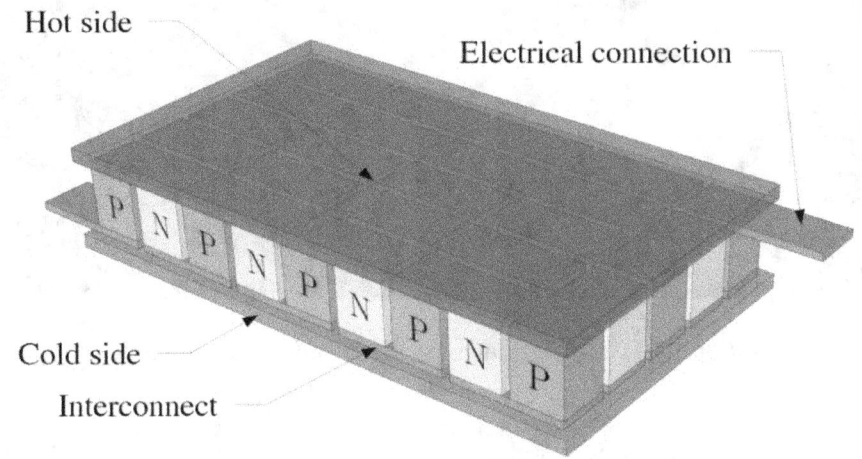

Figure 4.19. The structure of a Peltier cooler.

As shown in Figure 4.19, by forcing a current through the structure, a "hot side" and a "cold side" develop. Thus, the cold side can be used to cool something that is placed in contact with it.

As was described in Chapter 3, the "Peltier effect" and the "Seebeck effect" are essentially the same phenomenon; thus, this device could just as easily be called a "Seebeck cooler."

One obvious application of the Peltier cooler is to cool the CPU of computers and similar devices. A more practical device can be put to a more practical use.

Figure 4.20. A USB-port-powered Peltier cooler, designed to cool beverage cans (and their contents).

This Peltier cooler is powered using a USB port, and provides a Peltier cooling surface (the dark circle inside the white pedestal piece). This device provides a cooling surface, onto which can be placed beverage cans; the Peltier cooling surface is able to cool the contents of the beverage can.

Used in the opposite way, if a surface can be heated, a Peltier device can use the temperature difference between the heated surface and an unheated surface to generate electricity.

Of course, one of the core principles of thermoelectricity is its interchangeability – that electric power can create a temperature gradient (as in the Peltier cooler), *and* that a temperature gradient can produce electric power.

Space probes (such as earth-orbiting satellites) are able to create their needed on-board electric power via the use of solar panels; in the vicinity of the Earth, solar insolation is strong enough for a reasonable array of solar panels to be able to provide a power source. However, deep-space probes (such as the Voyager probes sent to the outer planets) are unable to make use of solar panels for power; in deeper space, sunlight is too weak for a reasonably-sized array of solar panels to be able to provide electrical power (and, of course, this situation worsens as a probe moves even deeper into space and further from the sun).

Instead, deep space probes create their on-board electricity by using fissionable material of some sort as the primary energy source. The decay of the fissionable material produces heat, the heat is used to heat one end of a Seebeck-like apparatus, and the temperature gradient thus produced is used to generate electrical power.

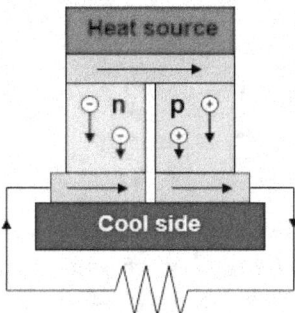

Figure 4.21. A Seebeck-effect-based thermocouple, which uses heat to generate electricity.

In this form, Seebeck's work is literally making possible the exploration of the universe.

The basic work on thermoelectricity by Seebeck was both scientifically fundamental – *and* the entry point for a number of practical applications at the intersection of heat and electricity.

Pondering Light

Seebeck's work on the behavior of chlorides of silver (when exposed to light) was to have implications in a complete different field.

As might have been obvious to the reader, when Seebeck showed that certain chlorides of silver were able to take on the color of the light to which they had been exposed, he had more-or-less invented color photography. Of course, there had been work on these materials prior to Seebeck's studies – as this excerpt from a 1927 retrospective makes clear.

> 298　　　JOURNAL OF CHEMICAL EDUCATION　　　MARCH, 1927
>
> ### THE CHEMISTRY OF PHOTOGRAPHY. I. HISTORICAL CONSIDERATIONS
>
> S. E. SHEPPARD, EASTMAN KODAK CO., ROCHESTER, NEW YORK
>
> In glancing backward toward the sources of the present broad stream of photographic evolution we encounter a fact common in the history of invention. It is impossible to rightly designate any one person as the inventor of photography, although attempts have been made to do so. Anyone who has ever traced a river to its "source" will remember how many little rills diverge as the stream is followed backward and upward, and how at last more than one spring or even damp patch of earth may be claimed as the one true source. No more easy is it to fix the exact source of a stream of invention. Of early discoveries leading to photography there might be mentioned the preparation of silver nitrate crystals (by the solution of metallic silver in nitric acid) described in the probably apocryphal writings of the Arabian alchemist, Geber. Although silver is the bedrock of modern photography, it has to spend much of its time

Figure 4.22. An excerpt from a 1927 paper reviewing some of the origins of the chemistry underlying photography. (S. Sheppard, Journal of Chemical Education, vol. 4, issue 3, p. 298.)

However, Seebeck's work had clearly laid down the more specific foundations of photography – and in particular, of color photography.

By the end of the nineteenth century, the first general-purpose, practical form of photography was introduced – more specifically, in 1888 by George Eastman.

Figure 4.23. George Eastman.

Eastman introduced a simple silver halide system of film that would take good-quality "black-and-white" (today known as grayscale) photographs. Eastman (and the company he founded, Kodak) made photography possible for non-specialists, by also introducing a simple camera – which contained the silver-halide-based film, and which was very easy to use.

Figure 4.23. An 1889 advertisement for Eastman's early Kodak camera.

In 1935, Kodak introduced a mass-market color film, Kodachrome. Kodachrome-based photography became the typical method of photography – right up until the recent rise-to-dominance of digital cameras and entirely-electronics-based photography.

Early Biochemistry

As described in Chapter 3, in 1815 both Seebeck and Biot noted that when polarized light was passed through certain aqueous solutions (such as sugar dissolved in water), the polarized light would emerge rotated from its original input polarization plane.

More details on this behavior – and its importance – can be explained by providing a larger version of an excerpt that was quoted in Chapter 3.

ROTATION OF THE PLANE OF POLARIZED LIGHT

Optically Active Substances. — It was known nearly a hundred years ago that when a beam of polarized light is passed through certain liquids, the plane of polarization is rotated or turned. This phenomenon was manifested by many substances in the crystalline condition, also by a number of carbon compounds in the liquid state and in solution. We are concerned here only with those optically active substances which are liquid at ordinary temperatures, or which are in solution. Some of the substances rotate the plane of polarization to the right and are called dextro-rotatory; others rotate to the left, and are termed lævo-rotatory. Dextro-rotation is indicated by the plus sign (+), lævo-rotation by the minus sign (−).

The number of substances whose rotatory power can be compared has increased enormously in the last few years. Biot and Seebeck pointed out in 1815 that certain organic substances have the power to rotate the plane of polarization. Oil of turpentine, and sugar and tartaric acid in aqueous solution, have this property, as was shown at this early date. From this time to 1879 the number of optically active substances increased to 300, while to-day we know over 700 substances[2] which have the power to rotate the plane of polarized light. The reason for the enormous activity in the preparation and

Figure 4.25. An excerpt from H. C. Jones, "The Elements of Physical Chemistry," 1902.

As this expanded excerpt explains, in the decades following Seebeck's work, more and more substances were found to exhibit this ability to rotate polarized light – and it was further observed that some substances rotated the polarized light one way, while other substances rotated the polarized light in the other direction.

As the excerpt notes, the substances in question (be they liquids on their own, or solids (such as sugar) that are dissolved in water) were "organic." As chemistry developed as a science in the nineteenth century, it underwent something of a division – into "organic chemistry" and "inorganic chemistry."

Originally, "organic" chemistry referred to materials which were connected to biology and living systems; eventually, because of the recognition that all known terrestrial life is based on carbon-centric molecules, the term "organic" was expanded to effectively refer to chemistry that involved carbon-based molecules. Everything else came to be known as "inorganic"

– though many simple molecules that contain carbon (such as carbon dioxide) appear on both sides of the created divide.

In contemporary terms, the field involving the very complicated carbon-based molecules more-directly associated with living systems has come to be termed "biochemistry," while the field involving simpler carbon-based molecules (whether related to living systems or not – in the latter case, for example, involving petroleum-based materials) is termed "organic chemistry."

However, during the nineteenth century, what would today be separated into "organic chemistry" and "biochemistry" were still taken to be of the same oeuvre. Of particular note, as the ability to carry out synthetic organic chemistry (that is, the ability to start with simple ingredients and then use chemical reactions to produce more complex outputs) was developed, it was found that it was possible to produce "synthetic" versions of already-known natural organic materials – including many that had previously been associated with biology.

This, of course, was recognized as being philosophically interesting – as it raised the simple question of, "What exactly *is* life?" Despite their different origins, the chemical behavior of the "natural" versions and the "synthetic" versions were identical – they were, materially, the same thing.

One interesting outcome of this situation was that the newly-developed processes for making paper produced, as a by-product, enormous amounts of vanilla extract – extract that was identical to natural vanilla extract, taken from naturally-grown vanilla. Vanilla is a tropical plant, and thus natural vanilla was, in most of the world, a scarce commodity. However, the ubiquitous growth of industrial-scale paper manufacturing produced (literally) as a by-product an equally-enormous amount of vanilla. As a result, vanilla became cheap, plentiful, and available everywhere. It is for this reason that vanilla-based food products became ubiquitous and inexpensive – which also caused the term "vanilla" to come to refer to the simplest base-case of nearly any effort to do something.

As noted in the 1902 excerpt of Figure 4.25, one of the substances that had been observed to rotate polarized light was tartaric acid. By the middle of the nineteenth century, organic chemists had been able to produce a "synthetic" version of tartaric acid. But this led to a bizarre inconsistency. While the "synthetic" tartaric acid was, in practice, exactly identical to "natural" tartaric acid – that is, "synthetic" and "natural" tartaric acid behaved in exactly the same way in chemical reactions – "synthetic" tartaric acid did not rotate polarized light.

This mystery was finally addressed via a difficult route in 1849 by Louis Pasteur – who, or course, would later go on to become famous for his other work.

Figure 4.26. Louis Pasteur.

Pasteur examined crystals of synthetic tartaric acid under a microscope – and he noted that some looked different than others; in fact, visually, he could see that by appearance, they could be divided into two separate groups.

Pasteur carefully sorted the crystals of tartaric acid into two groups based on their appearance – and when he had enough material in each group, he dissolved each group in water and tested the solutions with polarized light. Pasteur found a rather remarkable result. First, the solutions of the separated crystals of tartaric acid *did* rotate polarized light. And further, the two solutions rotated the polarized light in opposite directions. Finally, if the crystals are unsorted (or are sorted and then mixed back together), the resulting solution does *not* rotate polarized light – as had already been observed when there was no sorting or examination of the crystals.

What Pasteur had found forms what is more broadly considered to be the basis of biochemistry. Pasteur had found that some organic molecules – although "chemically" identical – exist in two different forms; these two forms rotate polarized light in opposite directions – and thus the two forms of the same material are known as *rotamers*.

In particular, these rotamers go beyond the simple idea of different molecular constructions using the same input atoms – beyond just simple assembly variations, there are complex organic molecules which exist in *two* (and only two) forms; as noted, these two forms are *chemically* identical-and-indistinguishable, but they rotate polarized light in opposite directions.

It was not understood at the time, but what Pasteur had found was the first example of what is known as *chirality* – the ability of a molecule to be structured around a central axis in either (effectively) a left-handed or a right-handed version.

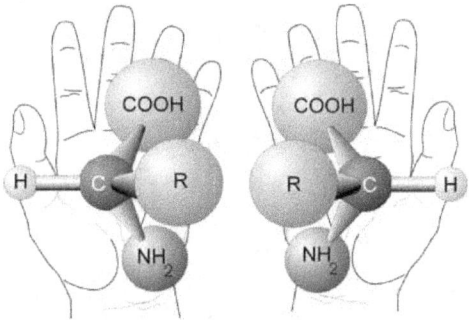

Figure 4.27. A simple example of chirality.

Complicated molecules – particularly those associated with complex biological functions – commonly showed this sort of chiral structure. And as the observation that "natural" tartaric acid rotates polarized light (and in one direction only) indicates, "natural" molecules often exist in only one of the two possible chiral forms.

A century after Pasteur's work with tartaric acid, the chirality of complex biological molecules was to emerge as a critical feature in the findings of James Watson and Francis Crick.

Figure 4.28. James Watson and Francis Crick.

Watson and Crick found that the DNA molecule – a very complex biological molecule that encodes and transmits the genetic characteristics associated with life – is structured with a twisting double-helix structure.

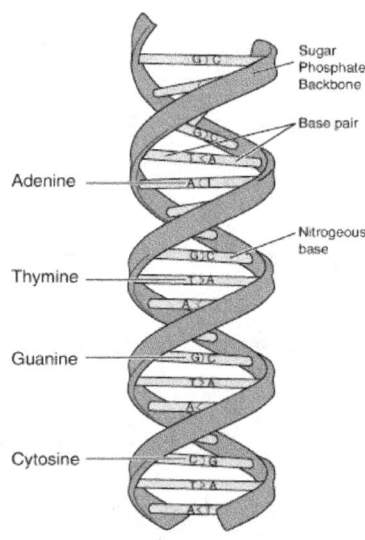

Figure 4.29. A detail (model) of a DNA molecule, showing the winding "double helix" of the outside chains of the molecule.

The complex DNA molecule could, in theory, be constructed using either of the two chiral arrangements of the winding outer double helix. However, in all life on Earth, the chirality is in one direction only. This shows the importance of the basic DNA structure to the evolution of life on Earth – as one of the two possible chiralities arose naturally at some point in the distant past, and then propagated (in that locked-in chirality) throughout time.

More recently, an interesting sidelight on the chirality of biological molecules has developed.

As was noted earlier, sugar was one of the materials which was found – when dissolved in water – to rotate polarized light. Sugar molecules ($C_6H_{12}O_6$) can, in theory, be constructed using either of two possible chiralities; however, like DNA, natural sugar molecules occur in only one of the two possible chiralities – as indicated by their ability to rotate polarized light.

Sugar molecules are, of course, readily broken down in cellular-level chemical reactions – via which they provide energy (and calories) to the host organism (such as… any of us). Not surprisingly, the human cellular system is able to process and use (i.e., metabolize) only those sugar molecules that possess the "natural" chirality direction; sugar molecules with the opposite chirality are unable to be processed by the cells.

In contrast, this situation does not hold with human taste receptacles – that is, the taste buds of the tongue. The taste buds are unable to distinguish between the two chiralities of the sugar molecule; to us, they both taste the same.

This has led to the development of zero-calorie sweeteners – based on sugar molecules with the "wrong" (i.e., not natural) molecular chirality. As the taste buds of the tongue are unable to tell the difference between the two chiralities, sugar molecules with the "wrong" chirality taste the same as those with the "right" chirality. However, when they reach the cells, these sugar molecules with the "wrong" chirality are unable to be processed to produce energy (or fat). At this level, the sugar molecules are biologically inert.

This is also an interesting contemporary consequence of some of Seebeck's original scientific work of the early-nineteenth century.

And perhaps one can ponder this by combining Seebeck's work in thermoelectricity with his work in the rotation of polarized light – by placing a can of a (zero-calorie) soft drink which uses the "wrong" chirality of the sugar molecule… onto a Peltier cooler, which keeps the beverage cool.

Final Thoughts

Isaac Newton supposedly stated that his achievements would not have been possible if it had not been for the efforts of those who had come before him – that "If I have seen further, it is because I stood on the shoulders of giants."

Today, we are in same situation. If we have seen further, it is because we are able to stand on the shoulders of a succession of giants.

Thomas Johann Seebeck certainly merits identification as one of those giants.

www.ingramcontent.com/pod-product-compliance
Lightning Source LLC
Chambersburg PA
CBHW071418220526
45469CB00004B/1322